Smart and Sustainable Manufacturing Systems for Industry 4.0

The current perspectives of smart and sustainable manufacturing systems hold important implications for current practices and understanding these concepts for further implications. This comprehensive reference text discusses both centralized and decentralized production systems using a variety of new cutting-edge approaches to solve problems.

The text covers simulation-based approaches, including social network-based approaches, discrete event-based approaches, and knowledge-based approaches for smart and sustainable systems. It further covers mathematical models such as single-objective, multi-objective, and many-objective. The text discusses important topics, including energy efficiency, transportation constraints for efficient and effective production, meta-heuristic and hybrid algorithms, and real-time monitoring and analysis for smart and sustainable production.

This book:

- Presents approaches to meet the objectives of sustainability and smart production systems.
- Discusses Internet of Things (IoT) and Industrial Internet of Things (IIoT) concepts and their implementation for production systems.
- Covers the social network analysis method in distributed manufacturing systems.
- Examines reckoning prognostics and diagnostics to monitor the health of systems from the perspective of distributed manufacturing.
- Discusses aspects of Industry 4.0 in specific production systems.

The text will be useful for graduate students and professionals in the fields of mechanical engineering, production engineering, industrial engineering, and manufacturing.

With the approaches highlighted in the book, the carbon footprint can be reduced to the highest extent, and the sustainable parameters of triple-bottom-line aspects will be greatly improved.

Smart and Sustainable Manufacturing Systems for Industry 4.0

Edited by
Vijaya Kumar Manupati,
Goran D. Putnik
Maria Leonilde Rocha Varela

CRC Press
Taylor & Francis Group
Boca Raton   London

CRC Press is an imprint of the
Taylor & Francis Group, an **informa** business

First edition published 2023
by CRC Press
6000 Broken Sound Parkway NW, Suite 300, Boca Raton, FL 33487–2742

and by CRC Press
4 Park Square, Milton Park, Abingdon, Oxon, OX14 4RN

CRC Press is an imprint of Taylor & Francis Group, LLC

ISBN: 978-0-367-64302-7 (hbk)
ISBN: 978-0-367-64303-4 (pbk)
ISBN: 978-1-003-12386-6 (ebk)

DOI: 10.1201/9781003123866

Typeset in Times
by Apex CoVantage, LLC

Contents

Preface

Recent paradigm shifts such as Industry 4.0, smart manufacturing, and China Manufacturing 2025 have changed the phases of production and manufacturing systems to a great extent. In other words, manufacturing has shifted from knowledge-based intelligent manufacturing to data- and knowledge-driven smart manufacturing. In order to convert the unprecedented volume of data into actionable and insightful information, there is a need for state-of-the-art technologies such as the Internet of Things (IoT), cyber-physical systems, artificial intelligence, machine learning, and deep learning techniques that act as catalysts for the previously mentioned new manufacturing paradigms. In addition to the mentioned technologies that cater to the needs of traditional approaches, achieving efficiency for today's customized, made-to-order, and decentralized environments requires a special kind of platform. Hence, besides the previously mentioned approaches, Industry 4.0 places emphasis on connecting digital technology, cyber-physical systems, and blockchain technology to achieve the need for connectedness and access to real-time processes, products, and people. Moreover, it is vital to investigate the approaches necessary to solve complex real-life problems using recent technologies. However, one must be aware of th other side of the coin, that is, challenges like security, sustainability, and proliferation of multi-modal data to overcome for better outcomes. However, the European Union has already taken concrete steps in the directions of controlling, coordinating, and communicating with triple-bottom-line aspects to meet the standards. Developing countries, particularly Brazil, Russia, India, and China (BRIC), have already taken part in this initiative.

Editors

Dr. Vijaya Kumar Manupati is currently working as an assistant professor in the Operations and Supply Chain, NITIE Mumbai. He received his Ph.D. in the Department of Industrial and Systems Engineering from the Indian Institute of Technology Kharagpur. His current research interests include intelligent manufacturing systems; agent/multi-agent/mobile-agent systems for distributed control, simulation, integration of process planning, and scheduling in manufacturing; sustainable supply chains; and evolutionary algorithms. He has more than 75 publications, which include journals like the *International Journal of Production Research, Computers, and Industrial Engineering*; *International Journal of Advanced Manufacturing Technology*; *Journal of Engineering*; *Journal of Measurements*; *International Journal of Computer Integrated Manufacturing*; and others. He acts as an international reviewer for more than 30 peer-reviewed journals. Currently, he is as an editorial review board member of the *International Journal of Sustainable Entrepreneurship and Corporate Social Responsibility*, IGI Global Publications. He received an Early Career Research Grant from the Department of Science and Technology (DST) for his research work on telefacturing systems. He is a member of the Institute of Industrial and Systems Engineering (IISE) and Institute of Engineers India (IEI) and a life member of the International Association of Engineers, US, and also acts as a technical committee member of various international conferences.

Maria Leonilde Rocha Varela received her Ph.D. in industrial engineering and management from the University of Minho, Portugal, in 2007. She is an assistant professor at the Department of Production and Systems of the University of Minho. Her main research interests are in manufacturing management, production planning and control, optimization, artificial intelligence, meta-heuristics, scheduling, web based systems, services, and technologies, mainly for supporting engineering and production management, collaborative networks, decision-making models, methods and systems, and virtual and distributed enterprises for Industry 4.0. She has published more than 200 refereed scientific papers in international conferences and international scientific books and journals indexed in the Web of Science and/or in the Scopus database. She coordinates R&D projects in the area of production and systems engineering concerning the development of web-based platforms and decision support models, methods, and systems.

Goran D. Putnik is Professor (Full), University of do Minho, Braga, Portugal. His education qualifications include a Diploma Engineer (five-year graduation course) in production engineering, Faculty of Mechanical Engineering, University of Belgrade, Belgrade, Serbia; M.Sc. in September 1988 (two-year postgraduate course) in production engineering, Faculty of Mechanical Engineering, University of Belgrade, Belgrade, Serbia; Dr.Sc. in January 1993 (three-year postgraduate course) in production engineering, Faculty of Mechanical Engineering, University of Belgrade, Belgrade, Serbia; the equivalency of Dr.Sc. in September 1994 by the University of

Mino, Braga, Portugal; and aggregation in March 2006: Aggregated and Dr. Habil., University of Minho, Braga, Portugal. He has served at various prestigious institutes as a professor at various levels and has more than 40 years of teaching experience as Assistant Professor (October 1995 to January 1999), Associate Professor (January 1999 to March 2006), Aggregated Professor (Associate Professor w/ Aggregation) (March 2006 to May 2009), and Professor (Full Professor) (June 2009 to date). He has authored more than 1000 articles, including international and national journals and various book chapters and books, and he has published in journals that include *CIRP Annals: Manufacturing Technology*; *International Journal of Production Research, Computers and Industrial Engineering*; *International Journal of Advanced Manufacturing Technology*; *Journal of Engineering*; *Journal of Measurements*; *International Journal of Computer Integrated Manufacturing*; and others. He facilitated the Prize "Literati Club 2005" Highly Commended Award presented to the authors Frans M van Eijnatten and Goran D. Putnik for the article "Chaos, Complexity, Learning and the Learning Organization: Towards a Chaordic Enterprise", *The Learning Organization: An International Journal*, Vol 11, No. 6, 2004. He acts as an international reviewer for peer-reviewed journals and is currently an editorial review board member for various journals.

Contributors

Alok Yadav
Associate Professor
MNIT Jaipur, India

Ana Araújo
Post-Doctoral Fellow
University of Minho, Portugal

Ana I. Pereira
Coordinate Professor
Polytechnic Institute of Bragança, Portugal

Ana Maria A. C. Rocha
Associate Professor
University of Minho, Portugal

Anbesh Jamwal
Research Scholar
MNIT Jaipur, India

Cátia Alves
Assistant Lecturer
University of Minho, Portugal

D.E.B. Costa
Researcher
University of Minho, Portugal

Dario Antonelli
Associate Professor
Politecnico di Torino, Italy

Fernando Romero
Assistant Professor
University of Minho, Portugal

Filipe Alves
Research Scholar
University of Minho, Portugal

Giulia Bruno
Ph.D. Scholar
Politecnico di Torino, Italy

Goran D. Putnik
Professor
University of Minho, Portugal

Guilherme Pereira
Associate Professor
University of Minho

Hélio Castro
Pos-doc Reseracher
University of Minho, Portugal

J.F. Lima
Resercher
University of Minho, Portugal

José Machado
Assistant Professor
University of Minho, Portugal

Leonel Patrício
Ph.D. student
University of Minho, Portugal

Lino A. Costa
Associate Professor
University of Minho, Portugal

Luís Dias
Assistant Professor
University of Minho, Portugal

Luís Ferreira
Post-doc Reseracher
University of Minho, Portugal

Luís Fonseca
Coordinate Professor
Polytechnic Institute of Porto, Portugal

Manuela Cunha
Professor
Polytechnic Institute of Cavado and Ave,
Portugal
Marcelo Henriques
Assistant Lecturer
University of Minho, Portugal

Margarida Pires
Researcher
University of Minho, Portugal

Maria Leonilde Rocha Varela
Assistant Professor
University of Minho, Portugal

Maria Pereira
Reseracher
University of Minho. Portugal

Marina A. Matos
Ph.D. Student
University of Minho, Portugal

Nuno Lopes
Professor
Polytechnic Institute of Cavado and Ave,
Portugal

Nuno O. Fernandes
Professor
Polytechnic Institute of Castelo Branco,
Portugal

Paulo Ávila
Coordinate Professor
Polytechnic Institute of Porto, Portugal
Paulo Leitão
Professor
Polytechnic Institute of Bragança,
Portugal

Rajeev Agrawal
Associate Professor
MNIT Jaipur, India

S.F. Moreira
Reseracher
University of Minho, Portugal

Sílvio Carmo-Silva
Professor
University of Minho, Portugal

Sónia Ribeiro
Researcher
University of Minho, Portugal

Vaibhav Shah
Invited Professor
University of Minho, Portugal

Vijaya Manupati
Assistant Professor
NITIE Mumbai, India

Zilda de Castro Silveira
Assistant Professor
University of São Paulo, Brasil

1 A Framework for Collaborative Practices Platforms for Humans and Machines in Industry 4.0–Oriented Smart and Sustainable Manufacturing Environments

Luís Ferreira, Goran D. Putnik, Maria Leonilde Rocha Varela, Vijaya K Manupati, Nuno Lopes, Manuela Cunha, Cátia Alves and Hélio Castro

CONTENTS

DOI: 10.1201/9781003123866-1

1.1 INTRODUCTION

Collaborative practices are of the utmost importance today in the context of Industry 4.0 and further to enable communities to reach sustainability (Campanella et al., 2012; Varela et al., 2014; Arrais-Castro et al., 2018; Varela et al., 2018, 2019; Putnik & Putnik, 2019). Collaboration is a term that is currently used and widespread through different contexts and for different purposes. Given distinct existing points of view and definitions arising from several organizations, entities and authors, a more or less general definition is provided by Mike Gotta,[1] where collaboration is defined as:

> social skills, relationships, practices and technology services that improve how people work jointly and substantially together (shared responsibility and risk) to communicate needs, coordinate activities, share information, exchange know-how, build community or achieve a common (team) objective (typically related to a process or project), within or across organizational boundaries.
>
> Although, collaboration can be further seen as an interaction between two or more entities or persons to enable to reach individual, not common goals (Putnik et al., 2021a, 2021 b) between individuals and/or organizations.

Sustainability is another emergent topic in new business models that is being reinforced with Industry 4.0. This happens today due to the central concern that exists currently about ecomimical, social, and environmental concerns in the dayly life of individuals and organizations. The impact of humans on manufacturing can now be more objective and focused, leaving mechanical and repetitive tasks for machines and processes but allocating supervision and cognitive decisions to humans. Indeed, the collaboration between all these agents (humans, machines and processes) is, as clearly evident and pertinent, an opportunity to explore. With the new advent of artificial intelligence (AI), the new perspectives arising from the Internet of Things (IoT), cloud computing (CC) maturity, and the high data processing and analysis capacity of big data and data science, new boosts to production levels and intelligent decisions are expected. Nevertheless, the impact of the symbiosis of both concepts, smart and sustainability, on manufacturing is not yet sufficiently clear. What is clear is the opportunity to explore: i) the increased processing capacity coming from smart and distributed cloud and edge computing; ii) the capacity to continuously monitor processes, persons and machines with IoT sensor networks; and iii) knowledge about the behavior of any production agent and cyber-physical system.

In this chapter, the main issue to be considered is the relevant aspects regarding collaborative practices for human and/or machine interactions in the context of Industry 4.0. For this purpose, it is important to start with a considered, proper definition of the pillars considered fundamental for supporting the proposed view regarding collaborative practices, along with a proposed framework for its support. In this regard, the proposed main pillars of the collaboration concept to promote

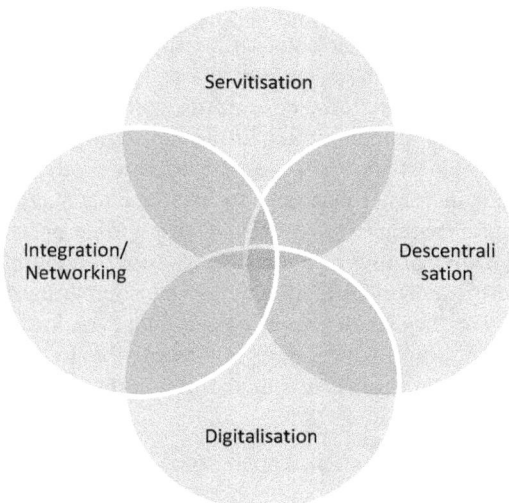

FIGURE 1.1   Main pillars of the proposed collaborative concept.

collaborative practices are: servitization, integration/networking, digitalization and decentralization (Figure 1.1).

Next, more general insights about collaboration are drawn, along with the identification of specific application domains regarding human and/or machine interventions, based on literature review, reflections and contributions. After this, a proposed framework for supporting collaborative practices is presented and briefly described. Next, the proposed framework is further analyzed through a comparison carried out with a set of five other existing frameworks regarding a set of 17 criteria that were considered important for establishing a comparative analysis between the frameworks. Finally, some main conclusions and future work intentions are presented.

1.2   A REVIEW OF COLLABORATIVE PRACTICES

Collaboration practices for humans and/or machines already have a long history, mainly arising from the context of Industry 3.0 (I3.0), as can be seen through several contributions that have been put forward during the last decades about these subjects, for instance, in Windeler & Sydow (2001) and Anumba (2002). Through this past history of collaborative practices, it is noticeable that the main underlying paradigms and technologies, which continue to be important and used today and which are also definitely of the utmost importance for the future in the forthcoming context of Industry 4.0 (I4.0), have already been put forward during the I3.0 period and even before, for instance, regarding generalized applications of artificial intelligence, along with more specific technologies and approaches, such as those based on neural networks, multi-agents and different kinds of machine learning mechanisms and algorithms (Anumba, 2002).

The main novelties or requirements concerning collaborative practices in the context of I4.0 are related to response/solution quality, on the one hand, and, on the other, to its availability regarding response time and capabilities to reach everyone/everything for each specific request concerning collaborative manufacturing and management processes or functions. Regarding the required high quality levels of responses/solutions, some of the main new contributions arising in the context of I4.0 are related to an increased necessity to use data science-related technologies and methodologies, for instance, for dealing with big data processing and analysis and those based on deep learning mechanisms. Moreover, a more generalized use of the cloud, along with exponential technologies and computing capabilities for fast responses on a real-time basis, are becoming feasible, even for highly intensive and complex problems to be solved, and can be made available to everyone, everywhere, based on fully digitalized machines and tools through the IoT.

1.2.1 COLLABORATIVE APPLICATION DOMAINS

Collaboration can be better contextualized regarding its underlying practice in terms of its application domains, and in this sense, we can differentiate, among other aspects, human-human, machine-machine and human-machine hybrid collaboration practices or domains. (Nunamaker & Briggs, 2010; Schuh et al., 2014). In the human-human application domain, there are collaborative practices just between humans, and important key words arising in this context of human collaboration domains are co-creation, just collaboration, pragmatics and smart collaboration-based human practices (Strohkorb et al., 2016), among others (He & Han, 2006). In the machine-machine application domain, there are collaborative practices just among machines (including system-system or machine-machine collaboration), and important key-words in this context of machine-machine or system-system collaborative practices are mostly related to multi-agent system (MAS)-based architectures, where interoperability issues and supporting middleware play a crucial role (Cheaib et al., 2011).

In the human-machine application domain, collaborative practices of hybrid or human-machine types occur. In this new world of cognitive computing, humans and machines work better together than individually by helping humans to expand the scale of their expertise to solve a whole new class of problems. In this way, a collaboration between human and machine boosts productivity, fosters new discoveries and frees up resources that could be put to better use doing tasks beyond the reach of purely digital systems (Robb et al., 2019). In this context, many occurrences are related to human–robot collaboration, and the expression "collaborative robots" is used. It is, therefore, a domain in which humans typically co-work with artificial intelligence-based systems and other machines rather than using them as tools. In this context, most successful collaborations involve bringing together abilities from one that the other lacks (Parodi & Gerio, 2017). Therefore, the main purpose of machine-human collaboration can be defined in the context of collaborating[2] to use the particular strengths of both types of intelligence, and even physical capabilities, to fill in for the other's weaknesses (Marvel & Norcross, 2017).

There are also many other possible contextualizations of the term "collaborative X", such as collaborative: communities, organizations, enterprises, factories, systems,[3] (work)spaces, environments, networks,[4] grids, platforms, frameworks, technologies,[5] machines/robots, software/groupware/tools,[6] businesses, services/tasks, projects, design manufacturing, maintenance, decision-making, methods[7]/methodology[8]/learning,[9] intelligence/thinking,[10] research,[11] and paradigms (DiRusso & Douglas, 2013), among others.

1.2.2 COLLABORATIVE PLATFORM FRAMEWORKS AND ARCHITECTURES

Software engineering continuously explores several modeling and implementing methodologies/technologies to align information systems with changing and emergent requirements or new capacities of arriving technologies (Figure 1.2). System architecture is a current tool that software engineers use to represent and describe this kind of system. Several domain-specific architectures were explored, and new architectures resulted from existing ones, composing and/or reorienting them. The business paradigm changed from a vertical orientation to today's globalization (Figure 1.2a), and information systems follow that path, too, from vertical and isolated applications toward recent distributed microservice architectures, an essential

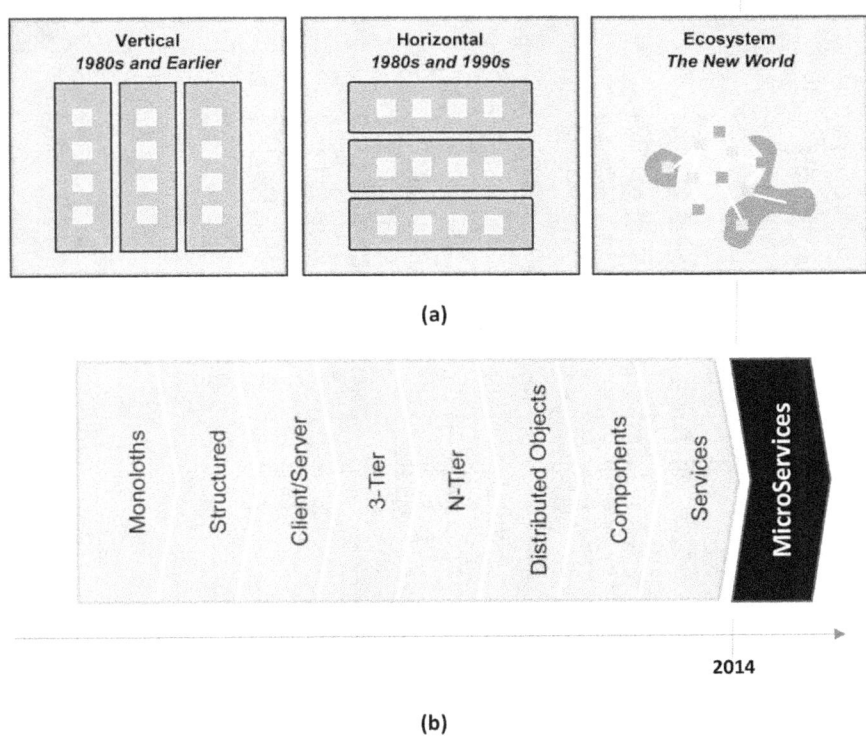

FIGURE 1.2   Business evolution and corresponding IT solution architectures.

Source: Adapted from Endrei et al., 2004

requirement for new cloud-oriented applications (Figure 1.2b). Collaboration is seen as a particular feature, and recent architectures try to represent and give directions to implement it technologically.

We assume that the cloud computing paradigm represents a relevant milestone in this process, making architectures appear focused on services that represent the required middleware to integrate isolated systems.

1.2.2.1 Services-Oriented Architecture–Reference Architecture

Services-oriented architecture–reference architecture (SOA-RA) (Yuan et al., 2016) is a consistent architecture for applications based on *services*—seen as models of real-world business activities, that is, mechanisms that allow the invocation of processes between different systems, supported on different platforms and implemented with different technologies.[12] Basically, it is a multi-layer architecture with well-identified and distributed responsibilities (Figure 1.3), where users only passively use the system, conditioning their context-aware autonomy and choice.

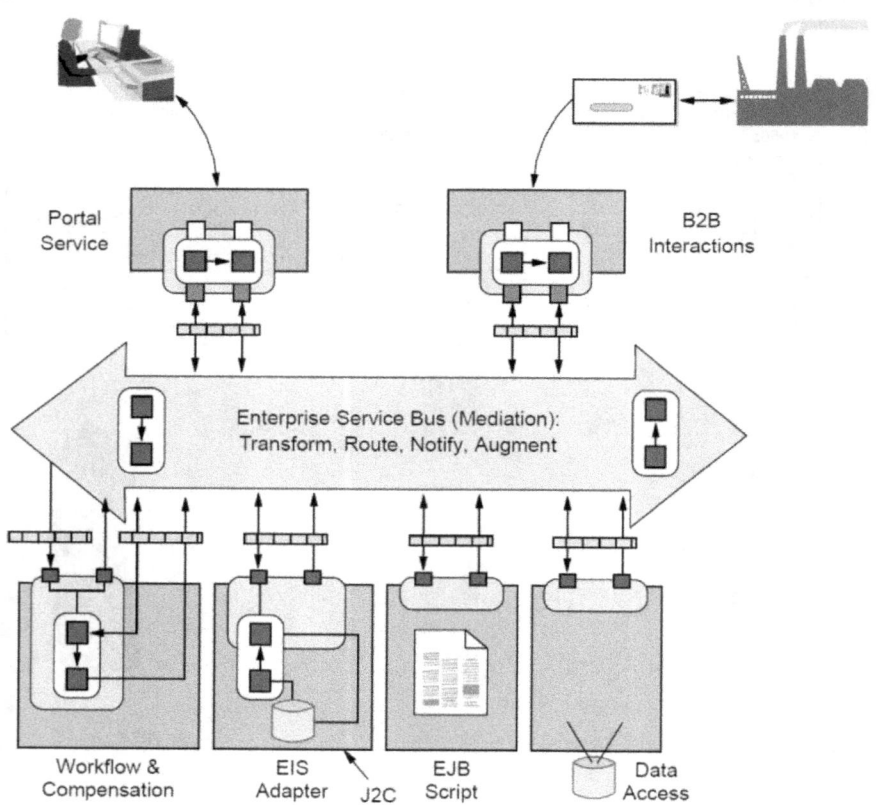

FIGURE 1.3 Layers of SOA architecture.

Source: Endrei et al., 2004

Besides the business agility and IT flexibility promised by services and their corresponding portability and loose coupling, these service-dependent strategies require complex processes for their orchestration and governance, as well as for the assurance of their quality of service (QoS) and security (Jensen et al., 2007).

1.2.2.2 Agile Architecture Framework

The *agile architecture framework* (AAF) is a very recent system architecture proposal for the expected digital enterprises: loosely coupled services fully aligned with business strategy, "architecture patterns that leverage the latest software innovations in distributed computing, autonomous systems, data streaming, and artificial intelligence" toward a "Boundaryless Information Flow vision achieved through global interoperability in a secure, reliable, and timely manner" (Barbazange et al., 2018). Organizational agility is the essential requirement, as the existing applications must be prepared to easily integrate new requirements, "supporting the active evolution of the design and architecture of a system while implementing new system capabilities". The software architect's role will change, for sure, moving from the traditional (nonexistent) project requirements and experience-based decisions to "focus on simple architecture that focuses on delivering customer value quickly" (Abrahamsson et al., 2010), learning and designing continuously when collecting field experience (practices, processes and knowledge) in a collaborative base, involving developers or advanced assisting tools and facing an *Architecture as a Service* (AaaS).

However, these scenarios of several dynamic components and required interoperability for the expected highly interconnected systems contribute to complex architectures and frameworks that are hardly implemented (Sturtevant, 2018). From another perspective, it is required that an

> enterprise's platforms have sufficient technical infrastructure to support the implementation of the highest-priority Features and Capabilities in the backlog without excessive redesign and delays . . . the enterprise must continually invest in extending existing platforms, as well as building and deploying the new platforms needed for evolving business requirements.
>
> (Leffingwell, 2011)

1.2.2.3 Internet of Things Architecture and Collaborative Internet of Things

The main concept around the IoT is the connectivity and interoperation of objects, people, systems and information resources. It is been explored in multiple areas such as health, industry or environment. In manufacturing, for instance, Wakefield (2014) states that the IoT will "eliminate massive information gaps about real-time conditions on the factory floor that have made it impossible to fully optimize production and eliminate waste in the past". The *Internet of Things Architecture* (IoT-A) is a proposal that essentially focuses on the task of supporting data gathering from physical device sensors and services to process it. Figure 1.4 shows an implementation of such architectures, in this case a smart city system, where several layers make a bridge from the IoT application to electronic devices.

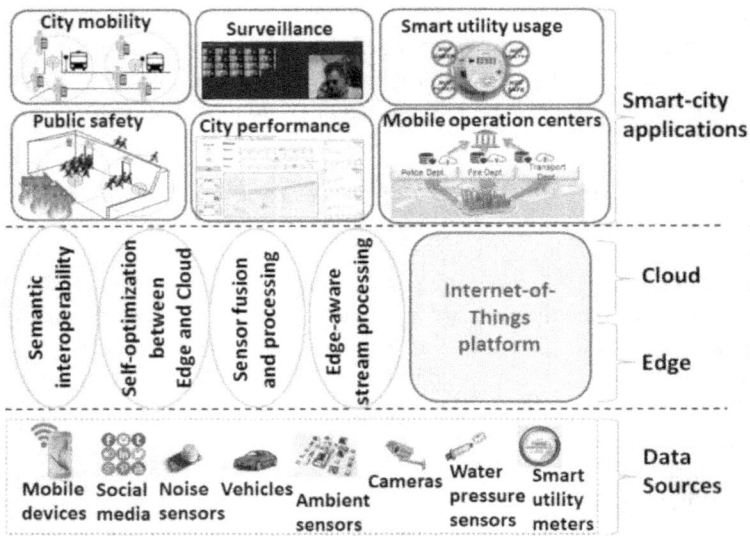

FIGURE 1.4   Smart city—example of IoT architecture implementation.

Source: IoT Team, 2016

The expected capacity to cooperate in data gathering (crowd sensing) and service sharing gave the *Collaborative Internet of Things* (C-IoT) the opportunity to emerge (Behmann & Wu, 2015). Collaboration here represents connectivity, processing and acting/reacting in the virtual or real world, mainly with electronic sensing devices. The majority of architectures that support C-IoT applications are based on common multi-layer architectures, with a particular layer dedicated to data acquisition from collaborating sensors/actuators (Figure 1.5).

Besides the IoT's architecture potential, the huge possibilities for implementing it and multiple entry points still present critical technological limitations, namely security, trust, privacy and identity management, data ownership and integrity, integrability, interoperability and composability, as well as law, social norms and culture, information fields and motivations as non-technical limitations (Ferreira et al., 2017).

1.2.2.4   *Cloud-Based Architecture for Ubiquitous and Cloud Manufacturing*

Ferreira et al. (2017) proposed a *cloud-based architecture for ubiquitous and cloud manufacturing* (CBUCM), where *pragmatics* appears as an innovative "tool" to allow active human participation in the system.

According to the authors, technology must be used only to model tangible things, and humans must support the remainder, resorting to communication tools (Figure 1.6) that support co-collaboration and effective co-decision. This architecture focuses on the effectiveness of the systems, a basic criterion for the expected dynamic reconfiguration when systems (including humans) and requirements are required to be aligned.

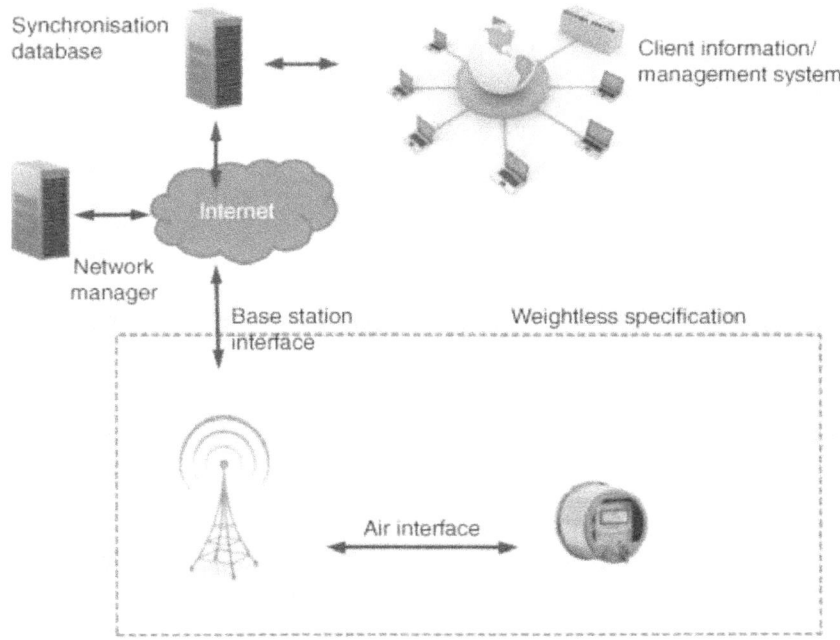

FIGURE 1.5 Example of C-IoT architecture.

Source: Ferreira et al., 2017

We assist this tendency to integrate communication "channels" that allows any-where, any way and anytime human-to-human collaboration (Skype, for instance). The *email* service is an example where the effectiveness can be conditioned . . . if an answer is required in the immediate context, later could be not so useful.

1.2.2.5 Remarks

Considering these architectures, one can easily deduce that each architecture is pre-pared to support a particular objective in a particular context. No architecture can easily support or substitute for others, and some try to integrate many others. We still continue with the axiom that "no one size fits all", as always.

1.3 FUNDAMENTALS FOR A COLLABORATIVE PLATFORM FRAMEWORK

Collaborative frameworks play a crucial role nowadays and are one of several import-ant collaborative X, being the key X further explored in the content of this special issue. To this end, some detailes will be provided in the context of I3.0 and I4.0.

> *Collaborative practices in I3.0*: Many different models, approaches, frame-works, architectures, platforms and tools have been widely put forward and

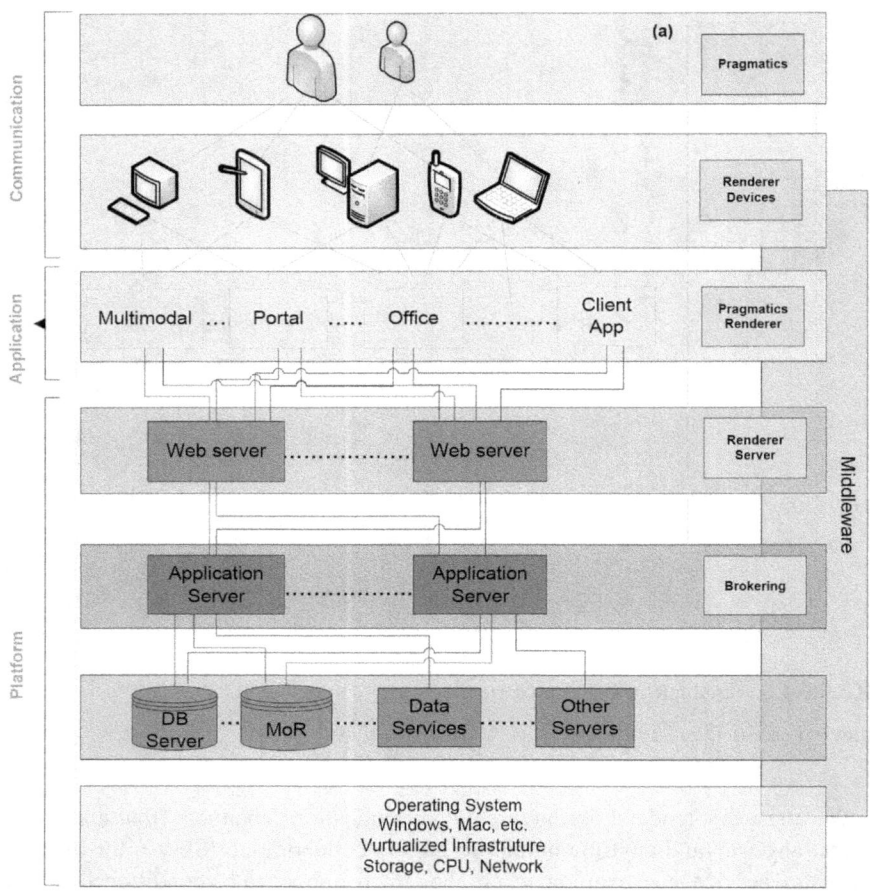

FIGURE 1.6 Architecture with pragmatics renderer.

Source: Ferreira et al., 2017

spread through several distinct scientific events and documents over many
years regarding more or less collaborative ones in the context of I3.0. Some
examples are thought to support collaborative work (Heinz-H, 2005).

Industry 4.0 or I4.0: The fourth industrial revolution is closely related to
the IoT, cyber-physical systems (CPSs), information communication technology and electronics (ICTE) and enterprise integration (EI) (Lu, 2017;
Gustavsson et al., 2017).

Collaborative practices in I4.0: Important contributions in the context of collaboration in the scope of I4.0 are related to quite different aspects and
domains, varying from collaborative frameworks (Johannes & Jivka, 2016;
Sipsas et al., 2016) to collaborative maintenance (Karakaya & Qi, 2014),
collaborative localizations in networks (Naitove, 2017), human–robot

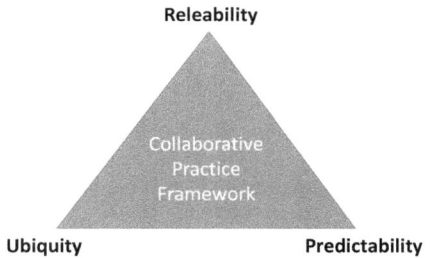

FIGURE 1.7   Main needs of a framework for collaborative practices.

collaboration (Nunamaker & Briggs, 2015), collaboration systems and technology (Wang et al., 2010) and multi-agent system-based collaboration (Cheaib et al., 2011), among other approaches and collaboration models, such as the example presented in Reardon et al. (2015) regarding the proposal of a machine-machine collaboration formalism based on web services.

Based on the literature review and also on our own developments, we emphasize the importance of three pillars as main requirements of a framework for supporting collaborative practices based on *reliability*, *predictability* and *ubiquity*, as schematized in Figure 1.7. *Reliability* refers to the capability of a framework to provide quality- and real-time–based responses or solutions in a fully interoperable way. *Predictability*, on the other hand, is related to the capability of a framework to anticipate fundamental actions, for instance, regarding the necessity of quickly changing some processing conditions or even a manufacturing layout configuration and/or regarding knowledge acquisition and/or processing to overcome a given difficulty or lack of information for proper decision-making process support. Finally, the *ubiquity* requirement is related to the necessity of providing responses/solutions everywhere, through digitalized means and technologies, for instance, based on IoT or IIoT, along with end-to end connections.

1.4   PROPOSED FRAMEWORK FOR A COLLABORATIVE PRACTICES PLATFORM FOR I4.0

The framework for a *collaborative practices platform for Industry 4.0* (CPPI4.0) scenario must support the most relevant requirements in this emergent context. Knowing that the weakest link conditions the whole chain's efficiency, in this scenario, it's required that there be superior alignment between enterprise elements (persons, infrastructures, systems, processes) and the global network of business, and all must interact with the capacity to communicate and collaborate. Accepting this in a purely social context and pushing technologies to support it when humans and machines interact and need to collaborate represents a huge challenge. The technology does not know everything! Collaboration represents data, communications and services that must be exchanged between all stakeholders. Their security, integrity, quality and trust are the keys needed for success. In a human-like world, cognitive collaboration is expected!

1.4.1 General Requirements

To better specify the expected supporting technological architecture, the expected use cases must be identified and analyzed. *Who, when, where, what* and *how* are known terms that indicate features and services to be supported. For example:

- The operator wants to know the history of the machine for which he/she is responsible. . .
- The operator needs to analyze the process state for which he/she is responsible. . .
- The coordinator needs "details" of the state of the unit that coordinates. . .
- The coordinator wants to interact immediately with any technician for the unit that coordinates. . .
- The coordinator wants immediate help in solving any special issue. . .
- The coordinator wants to be alerted in advance of an imminent problem. . .
- The coordinator wants to control rules/events for notification. . .
- The manager wants to manage his team performance: learning, creativity, adapting to change. . .
- The customer wants to monitor the state of his/her product. . .
- The customer wants to certify the product that he/she buys. . .

As anyone can easily infer from these and many other possible use stories, the capacity to get or set related data or information in a timely manner for any existing system, infrastructure, process or even person can be a determinant of the correct decision. These aspects are crucial to consider when designing the logical architecture of a platform for collaborative practices for humans and machines in the context of I4.0.

1.4.2 Logical Architecture

Cyber-physical systems (Lu, 2017; Gustavsson et al., 2017; Lee, 2008; Broy, 2010) are a symbiotic relation between computation and physical components (or processes) that must be understood as a dynamic and circular integration (Lee, 2008) where the ends can condition the beginnings. A physical product (or part of it) results from the initial "rules"—construction processes implemented, controlled and monitored by (digital or not) systems (which could be persons). The innovation comes from the possible realignment of the initial rules after in-course results are interpreted.

CPSs represent the paradigm for supporting platforms in I4.0 contexts, where reliability and predictability are mandatory criteria for the required integration of all involved components. Following standards, patterns and proven algorithms (in general computing engineering) is the key to success only if the environment is completely controlled. However, if the software is distributed across several (not necessarily controlled) processing entities, success can be postponed, and the same happens with physical components. Following engineering fundamentals, the components can be assembleable (composable) to work efficiently, but each component can fail, and global success is pushed, too. The (un)known complexity of merging these two worlds, computational and physical, represents the main challenge for a successful collaboration in I4.0. Accepting the intrinsic concurrency and inevitable consequences of time passing (Lee, 2008), these systems must efficiently deal with uncertainty and be sufficiently robust against unexpected asserts and subsystem failures. The technology (not necessarily digital) must be pushed to the limit toward the creation of reliable and predictable systems independently of its abstraction level. However, because not everything is technologically supported, intangible domains are not supported by technology (Ferreira et al., 2017), and a new level of abstraction must be considered to compensate for this and ensure robustness. This new level definitely must leave room for human decisions.

With the acceptance, alignment and integration of cloud services, the IoT represents a new kernel for global realignment and adaptation for the emergent business context where

> the market demand is moving from reactive on-demand remote monitoring and control (pull model) to more proactive services where the services will identify who you are and deliver what you want and when you want. Services will take on a new form empowered by insights driven from data collected in a given target applications.
>
> (Behmann & Wu, 2015)

In I4.0 and CPSs (Lu, 2017; Gustavsson et al., 2017; Lee, 2008; Broy, 2010), almost everything is expected to have an active part on this new scenario of the Internet of Everything (IoE), where persons and machines both belong, creating a digital life focused on quality, efficiency and productivity. A Collaborative Internet of Things (C-IoT) is expected, and human, industry and infrastructure will be affected by its impact, working toward a smart society with smart business and improvement of quality of life (Figure 1.8).

Thus, technology (computation and physics) and humans are the main components of the logical architecture that supports the I40 collaborative framework. Global collaborative capacity is essential, and effective human-machine, machine-machine and human-human relations are demanded, too. The "traditional" architecture model relies on sensing, gateways and services (Figure 1.8) where distinct domains (health, industry, energy, etc.) have their particular and non-interoperable implementations (Figure 1.9a).

It is expected now that any industrial application may span several domains. Personal sensing with smart wearables (human), along with operator health monitoring, will certainly be one input for machine efficiency (industrial) (Figure 1.9b)! Considering such built-in crowd sensing and adding human direct communication, collaboration and co-decision support, we are facing the effective collaborative practices framework (Figure 1.9c).

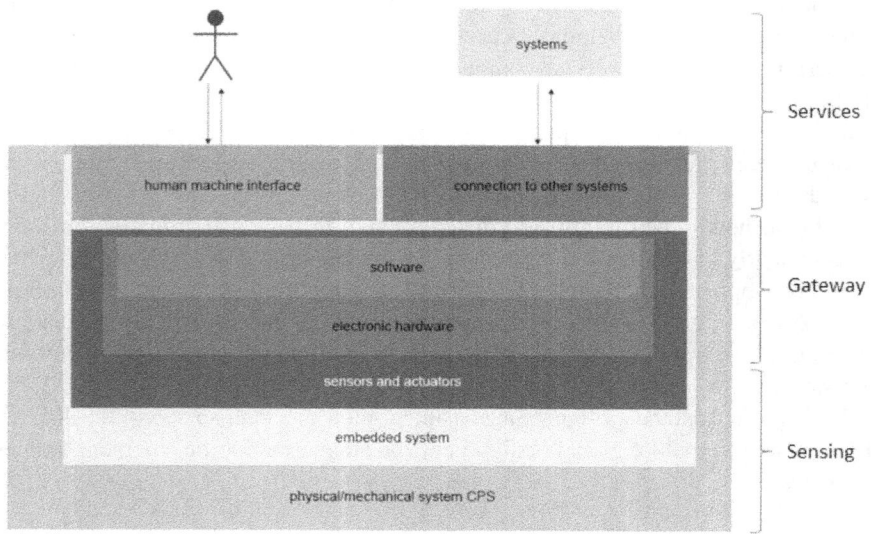

FIGURE 1.8 "Traditional" architecture model for supporting machine-machine and machine-human interactions for enabling collaboration in the IoT environment.

Source: Adapted from Broy, 2010

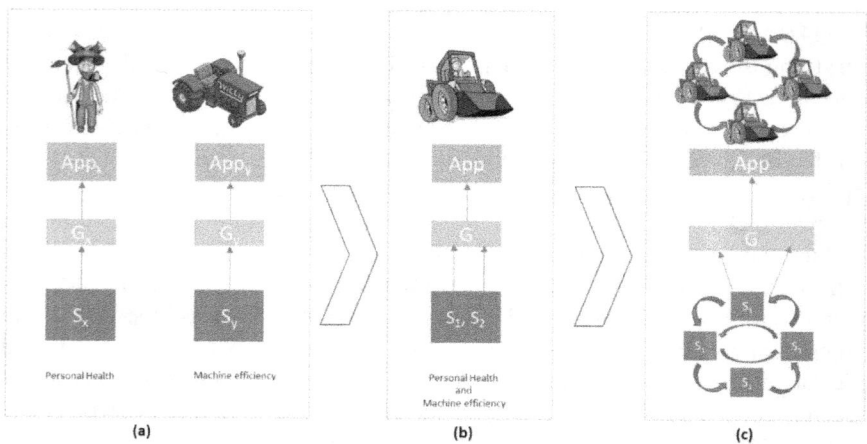

FIGURE 1.9 Toward the CPPI4.0 framework.

1.4.3 FUNCTIONAL ARCHITECTURE

A generic use case for the CPPI4.0 can be seen when a generic user needs to interact, directly or in a mediated way, with another user(s) or with a (possibly remote) *physical entity* (PE) in the physical world. The proposed architecture derives from

CBUCM and IoT-A architectures (described previously). Considering the possible usage scenario, four main abstractions were identified on the domain model: *users*—a human person or digital artifact (e.g., a *service*, a mobile app, a software agent, etc.); *virtual entities*—digital artifacts with characteristics of physical entities (e.g., object, avatar, database entry, social account, machine); *devices* that mediate interactions between physical and virtual entities (sensors, tags, actuators); and *resources*—software components (e.g., microservices, data) that run or flow on devices or cloud nodes (Figure 1.10).

Considering these entities, the functional architecture for CPPI4.0 (Figure 1.11) is structured from several integrated main components whose representation follows the basics of component-based framework software engineering (CBSE) notation (Bachman et al., 2020; Crnkovic & Larsson, 2002).

The *IoT process management* component is responsible for managing the execution of all possible interrelations with physical entities (data from sensors, orders to actuators, etc.), that is, combining business objects and processes with IoT services.

The *service organization* component is responsible for the composition and orchestration of service invocation/subscription and execution. It behaves as a hub for finding, selecting and using of any kind of functional service. Virtual entities (VE) and IoT services are abstraction layers that allow applications to interact with physical components or other users. With the physical devices' resources (e.g., temperature sensor) exposed as IoT services, it is necessary to discover that sensor and communicate its raw value. Thus, the virtual entity component, behaving as an entity

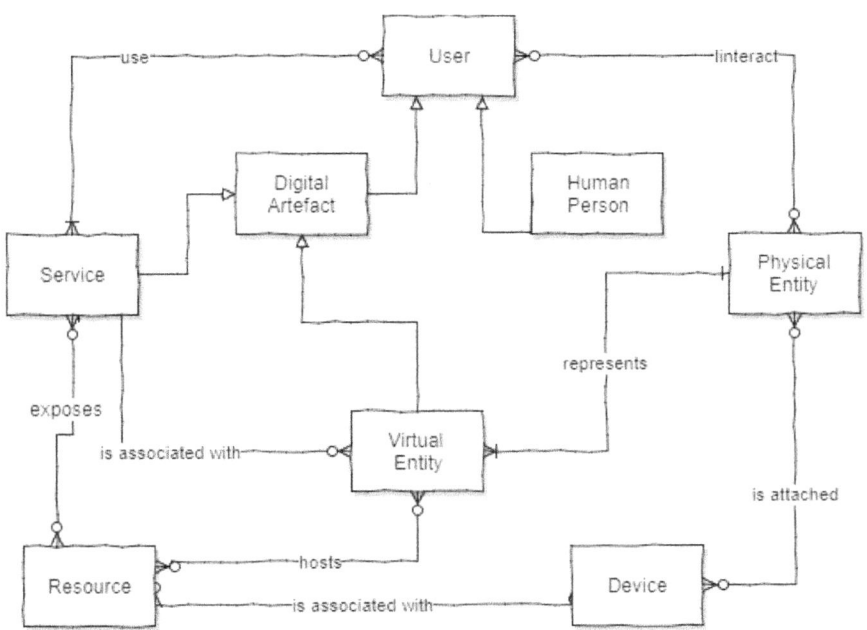

FIGURE 1.10 Main domain entity relations.

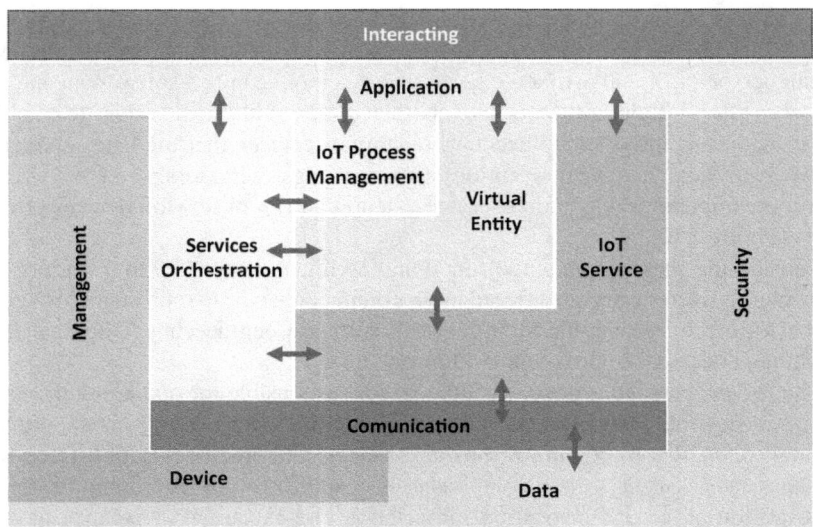

FIGURE 1.11   CPPI4.0 functional architecture.

broker, is responsible for finding and selecting the appropriate physical representation (virtual entity), as well as functionalities to compose and validate existing
associations or relations.

The *IoT service* component behaves as an IoT device broker. It has services to
interact with IoT devices, as well as dynamic brokering functionalities to find appropriate IoT devices for VE requests. The *communication* component is an essential
layer for any distributed system that allows multiple-protocol interoperability. Several
distinct electronic devices will be connected, and several IoT communications protocols (NFC, Zigbee, Wi-Fi, LoRa, etc.) will be integrated, exploring domain-specific
middleware and services. The *security* component ensures authorization, privacy,
trust and integrity in data, communications and processes realized between users
and data producer/consumer entities. The *management* component is responsible for
QoS management, device cataloguing and classification, rule definition and other
global management assets. Properties such accuracy, performance and robustness
will be handled in this component.

The *application* component consists of all possible digital artifacts (applications)
that will be integrated and use the platform. Dashboards for remote sensing machines
and remote control, alarm apps, wearable hubs and machine device care (MDC)
are some examples of applications. The *device* component essentially represents
the ground electronic devices or production machines, active or passive, that, producing or consuming data, execute a particular action. The *data* component represents any other repository (database, file, process) that can have or produce relevant
data. Finally, the *interacting* component represents the most innovative contribution for this framework, the collaborative practices set of services. It allows human
users to directly collaborate and co-decide in a timely way. Human-centered data

visualization and analysis tools, recurring to high processing data algorithms, virtual and augmented reality (VR/AR) and emergent artificial intelligence processes, will allow human cognitive sharing, immersive interaction, simulation and predicting scenarios. The overall platform will ensure collective impact, vigilance, learning and action in the social-like pattern that we intend to ensure.

The main technological supporting details follow.

1.4.4 TECHNOLOGICAL ARCHITECTURE

Globally, the proposed architecture models a distributed platform where:

1 Several distinct data sources are supported, processed, integrated and stored in a federated distributed repository.
2 Several services support: a) learning processes, applying pattern recognition and processing to real-time acquired data; b) high data processing and analysis; c) integration middleware for highly interconnected processes and systems; and d) advanced data visualization for governance and process monitoring.
3 A federated social network and pragmatics renderer support human-to-human collaboration for co-decision between federated members.

The technological platform is based on open source technologies and cloud services, including infrastructure as a service and platform as a service model for overall dynamic management services and crowd-sensing resources for an industrial raw data acquiring system (Figure 1.12).

It follows an N-tier architecture model with:

1 A *data acquisition layer* to acquire near–real-time data coming from electronic, digital or nondigital devices (sensors, actuators, tags, machines, wearables); from raw or structured components (databases, files); or from active system components (processes, persons).
2 A *data processing layer* to extract, analyze, process and store the acquired data and inferred knowledge.
3 An *application layer*, exploring application services.
4 An *interacting layer* for effective visualization, collaboration and actuation services on any platform (Figure 1.12).

All resources (processes, machines, persons, applications. services, etc.) must be appropriately described and registered. The brokers will look for them using specific ontologies, functional signatures and even intelligent pattern-matching processes. Big data, artificial intelligence, high-performance computing and human moderation shall be explored in those brokering processes, which will include finding services, gathering data and making decisions. Collaboration, synchronous or not, will be possible between any registered entity or resource, and middleware-oriented messages will ensure interoperability between services and agents. Through such messages will be possible to explore edge computing on extended IoT devices to manage their behavior and intercollaboration.

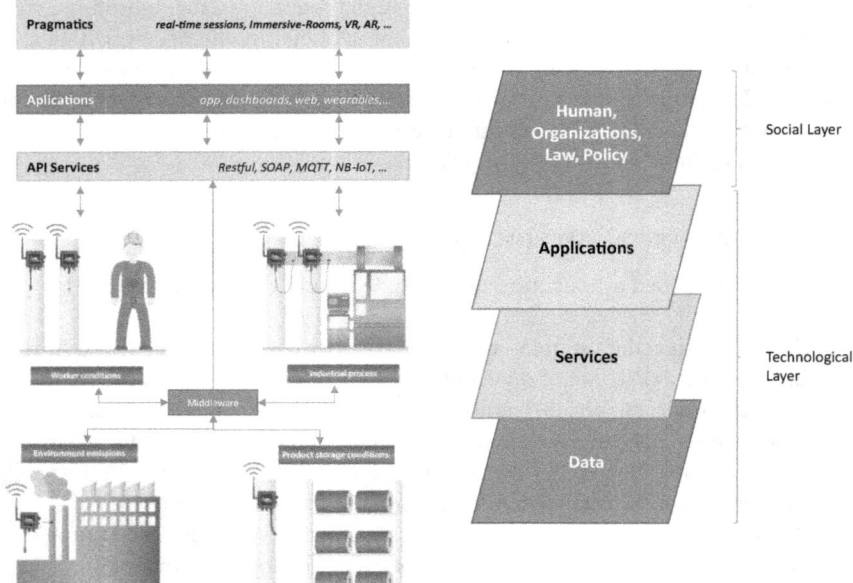

FIGURE 1.12   CPPI4.0 framework.

Source: Adapted from Libelium, 2015

The expected dynamic behavior of these activities will require the capacity to find continuously fitting resources. Indeed, the generative capacity to have predictive scenarios will be required, too. Visualization and interaction are also critical supporting services. The capacity to get the data, gather the existing knowledge on it and visualize and navigate effectively in all existing dimensions of that knowledge or inferred new knowledge represents the needed advantage of such collaborative platforms. Supporting decisions based on predictive scenarios or simulations are very relevant but don't have the capacity to use information favorably and in a timely way, implying an inability to learn and avoid problems. Immersive technologies (virtual and augmented reality), communications channels, collaboration services, event-based pushing services, cognitive services and microservices will represent the main spectrum of technological implementation (Figure 1.13).

1.5   I4.0 PLATFORM COMPARISON

After the previously described exposed architecture layers and functional components, Industry 4.0 and expected collaboration practices architectures must be smart and secure considering the following criteria: connectivity, pricing, high processing

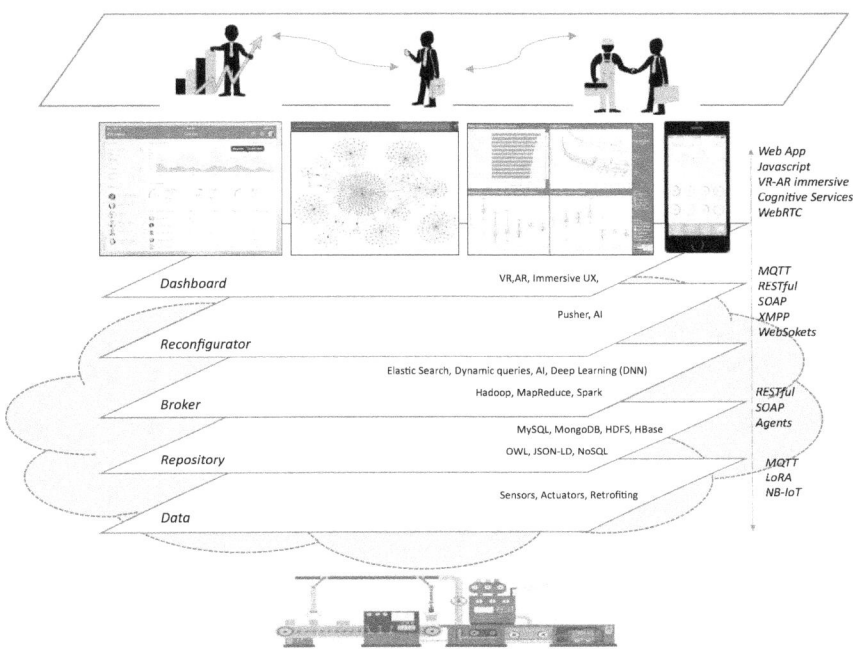

FIGURE 1.13 CPPI40 full technological stack proposal.

Source: Adapted from Ferreira et al., 2017

capacity and scalability, data science support, interoperability, memory (artificial intelligence) and multi-storage format, sensing, device control, pragmatics, trust, robustness and threat and identity management. Let us consider a typical industrial use case where industry users (technicians, managers, engineers, leaders) and related stakeholders (providers, costumers, others) will have a system implemented according to different frameworks or architectures. Table 1.1 shows the main differences of the explored architectures and criteria. In the table, "+" indicates the feature is implemented, "−" means the feature does not exist and "+−" means the feature either does not exist or is not intended for that purpose.

Through analysis of Table 1.1, it is possible to see that only our proposed platform architecture (CPPI4.0), within the set of six analyzed platforms, is designed to support all 17 of the listed criteria (C1 to C17). Next in the ranking is the CBUCM platform architecture, which meets 15 of the 17 criteria considered. After that is the IoT-A/C-IoT platform architecture, which meets 13 of the 17 listed criteria, followed by the AAF platform architecture, with 12 criteria satisfied. SOA-RA appears in the penultimate position, with 11 criteria met, and finally, the B-SOA platform's architecture only supports 5 criteria out of the set of 17 analyzed.

TABLE 1.1

I4.0 Architecture Comparison

Main Criteria	B-SOA[13]	SOA-RA	AAF	IoT-A/C-IoT	CBUCM	CPPI4.0
C1. Connectivity	–	+	+	+	+–	+
C2. Processing—Analytics	–	+–	+–	+–	+–	+–
C3. Processing—Dynamic Composition	–	+–	+	+	+–	+
C4. Processing—Scalability	–	+–	+	+	+–	+
C5. Memory—AI	–	–	+–	+–	+–	+–
C6. Sensing	–	–	–	+	–	+
C7. Security—Trust	+–	+–	+–	+–	+–	+–
C8. Security—Industrial Management	+–	+–	+–	+–	+–	+–
C9. Security—Threat Analytics	+–	+–	+–	+–	+–	+–
C10. Interoperability	–	+	+	+–	+	+
C11. Pricing	+–	+–	+–	+–	+	+
C12. Data Science	–	–	–	–	–	+–
C13. Pragmatics	–	–	–	–	+	+
C14. Context-Aware	–	–	–	–	+	+
C15. Reconfiguration	–	+–	+	+	+	+
C16. Open Source	+–	+–	+–	+–	+	+
C17. Collaboration	–	–	–	–	+	+
Total	5	11	12	13	15	17

1.6   CONCLUSION

Collaborative manufacturing and management is a key issue in the context of a sustainable Industry 4.0, along with the emerging new paradigms and practices arising in the underlying cyber-physical systems. The emergence of new supporting technologies for communications, processing and analysis support, together with new, increasingly able physical devices (sensors, controllers, actuators) and innovative systems to explore them has made smart and sustainable attributes emergent and relevant topics in manufacturing business models. The capacity to have multiple agents working together, that is, cooperating and, mainly, collaborating, requires the existence of dynamic and efficient interoperability practices. One of those practices is the in-time and continuous collaboration between all those expected manufacturing agents, namely humans, machines and processes. Collaboration can be connectivity or systems integration from a technical perspective or effective communication between humans and machines or between only humans.

In this chapter, the main pillars of a proposed collaborative concept were presented in light of collaborative practices platforms for Industry 4.0. Moreover, the main requirements of a framework for collaborative practices were also highlighted and different collaborative application domains described. The framework for a collaborative practices platform for I4.0 was proposed and described and a comparative analysis carried out with a set of other relevant existing platforms based on a set of 17 defined criteria. Through the description and the comparison, it was possible to see the value added by the proposed collaborative practices platform. Planned future work includes the complete development and implementation of the proposed platform to be further applied in real-life manufacturing application scenarios. The W3C Web of Things architecture recommendation represents an important step for application integration support, and an integration hub should be developed to gather data from distinct and isolated systems.

ACKNOWLEDGMENTS

This work has been supported by FCT—Fundação para a Ciência e Tecnologia within the R&D Units Project Scope: UIDB/00319/2020 and EXPL/EME-SIS/1224/2021.

NOTES

1 http://mikeg.typepad.com/perceptions/2004/05/defining_collab.html
2 http://whatis.techtarget.com/definition/machine-human-collaborationis
3 https://www.linkedin.com/pulse/collaboration-40-system-alberto-garcia-jurado; http://www.computerbusinessresearch.com/Home/enterprise-resource-planning-and-collaborative-systems/collaboration-system
4 https://en.wikipedia.org/wiki/Collaborative_network
5 https://www.igi-global.com/dictionary/collaborative-technologies/4436
6 https://en.wikipedia.org/wiki/Collaborative_software
7 https://en.wikipedia.org/wiki/Collaborative_method
8 https://www.scie.org.uk/publications/guides/guide34/background/whatis.asp
9 https://link.springer.com/referenceworkentry/10.1007%2F978-1-4419-1428-6_910
10 https://www.amazon.com/Collaborative-Intelligence-Thinking-People-Differently/dp/0812994906
11 https://ccrec.ucsc.edu/center/what-collaborative-research
12 http://www.soapatterns.org/enterprise_service_bus.php
13 Before SOA Architectures.

REFERENCES

Abrahamsson, P., Babar, M. A., & Kruchten, P. (2010). Agility and architecture: Can they coexist? *IEEE Software*, 27(2), 16–22.

Alberto Garcia-Jurado, CEO, Coach, Speaker and Innovator on Smart Collaboration and Cultural Intelligence. *Collaboration 5.0: The Collaboration System.* https://www.linkedin.com/pulse/collaboration-40-system-alberto-garcia-jurado.

Anumba, C. J., Ugwu, O. O., Newnham, L., & Thorpe, A. (2002). Collaborative design of structures using intelligent agents. *Automation in Construction*, 11(1), 89–103.

Arrais-Castro, A., Varela, M. L. R., Putnik, G. D., Ribeiro, R. A., Machado, J., & Ferreira, L. (2018). Collaborative framework for virtual organisation synthesis based on a dynamic multi-criteria decision model. *International Journal of Computer Integrated Manufacturing*, 1–12. Taylor & Francis.

Barbazange, H., Générale, S., Bunouf, J., Générale, S., Le, J., Générale, S., . . Polytechnic, E. (2018). *Agile Architecture in the Digital Age*. The Open Group.

Behmann, F., & Wu, K. (2015). *Collaborative Internet of Things (C-IoT): For Future Smart Connected Life and Business*. Wiley.

Broy, M. (2010). Cyber-physical systems innovation durch software-intensive eingebettet Systeme. *Acatech Diskutiert*, 1–141.

Campanella, G., Pereira, A., Ribeiro, R. A., & Varela, M. L. R. (2012). Collaborative dynamic decision making: A case study from B2B supplier selection. In *Decision Support Systems— Collaborative Models and Approaches in Real Environments*, 88–102. Springer.

Cheaib, N., Otmane, S., & Mallem, M. (2011). A machine-machine collaboration formalism based on web services for groupware tailorability. *Proceedings of the 2011 15th International Conference on Computer Supported Cooperative Work in Design (CSCWD)*, Lausanne, Switzerland, 238–245.

Crnkovic, I., & Larsson, M. (Eds.). (2002). *Basic Concepts in CBSE*. In *Building Reliable Component-Based Software Systems*. Artech House.

DiRusso, D. J., & Douglas, M. (2013). The validity of the technology acceptance model in collaboration systems software. *Business and Management Review*, 3(03), 01–05 May. ISSN 2047-0398

Endrei, M., Ang, J., Arsanjani, A., Chua, S., Comte, P., Krogdahl, P., & Luo, M. (2004). *Patterns: Service Oriented Architecture and Web Services*. RedBooks IBM.

Ferreira, L., Putnik, G., Cunha, M. M. C., Putnik, Z., Castro, H., Alves, C., . . . Varela, L. (2017). A cloud-based architecture with embedded pragmatics renderer for ubiquitous and cloud manufacturing. *International Journal of Computer Integrated Manufacturing*, 30(4–5), 483–500,

Francesco, P., & Paolo, G. G. (2017). AURA: An example of collaborative robot for automotive and general industry applications. *Procedia Manufacturing*, 11, 338–345. ISSN 2351-9789.

Giret, A., Trentesaux, D., Salido, M. A., Garcia, E., & Adam, E. (2017). A holonic multi-agent methodology to design sustainable intelligent manufacturing control systems. *Journal of Cleaner Production*, 167, 1370–1386.

Gustavsson, P., Syberfeldt, A., Brewster, R., & Wang, L. (2017). Human-robot collaboration demonstrator combining speech recognition and haptic control. *Procedia CIRP (The 50th CIRP Conference on Manufacturing Systems)*, 63, 396–401.

He, F., & Han, S. (2006). A method and tool for human–human interaction and instant collaboration in CSCW-based CAD. *Computers in Industry*, 57(8–9), 740–751. ISSN 0166-3615.

Heinz-H, E. (2005). Human-human collaboration. *2005 IFAC. 16th Triennial World Congress, Prague, Czech Republic*, 38.1, 72–77.

Herter, J., & Jivka, O. (2016). A model based visualization framework for cross discipline collaboration in Industry 4.0 scenarios. In *Factories of the Future in the Digital Environment— Proceedings of the 49th CIRP Conference on Manufacturing Systems. Procedia CIRP*, 57, 398–403. Elsevier.

IoT Team, 2020 Project. (2016). *IoT 2020: Smart and Secure IoT Platform*. IEC— Internaational Electrotechical Commission.

Jensen, M., Gruschka, N., Herkenhoner, R., & Luttenberger, N. (2007). SOA and web services: New technologies, new standards—new attacks. In *Fifth European Conference on Web Services (ECOWS'07)*, 35–44. IEEE.

Karakaya, M., & Qi, H. (2014). Collaborative localization in visual sensor networks. *ACM Transactions on Sensor Networks (TOSN)*, 10(2), 18.

Lee, E. A. (2008). Cyber physical systems: Design challenges. *2008 11th IEEE International Symposium on Object and Component-Oriented Real-Time Distributed Computing (ISORC)*, 363–369.

Leffingwell, D. (2011). *Agile Software Requirements: Lean Requirements Practices for Teams, Programs, and the Enterprise*. Addison-Wesley Professional.

Libelium. (2015). *Smart Factory: Reducing Maintenance Costs and Ensuring Quality in the Manufacturing Process*. https://www.libelium.com/libeliumworld/success-stories/smart-factory-reducing-maintenance-costs-ensuring-quality-manufacturing-process/

Lu, Y. (2017). Industry 4.0: A survey on technologies, applications and open research issues. *Journal of Industrial Information Integration*, 6, 1–10.

Marvel, J. A., & Norcross, R. (2017). Implementing speed and separation monitoring in collaborative robot workcells. *Robotics and Computer-Integrated Manufacturing*, 44, 144–155. ISSN 0736-5845,

Naitove, M. (2017). Connectivity & collaboration in robotics & automation at K 2016: Besides a handful of new robots and pickers, the big themes were modular 'plug-and-play' automation, Industry 4.0 connectivity, easier programming, and safer collaboration with human workers. *Plastics Technology*, 63(3), March, 12, 4 p.

Nunamaker, J. F., & Briggs, R. O. (2010). Introduction to collaboration systems and technology track. *Proceedings of the 41st Annual Hawaii International Conference on System Sciences*. IEEE.

Nunamaker, J. F., & Briggs, R. O. (2015). Introduction to collaboration systems and technology track. In *2015 48th Hawaii International Conference on System Sciences*. IEEE Xplore.

Putnik, G. D., & Putnik, Z. (2019). Defining sequential engineering (SeqE), simultaneous engineering (SE), concurrent engineering (CE) and collaborative engineering (ColE): On similarities and differences. *Procedia CIRP*, 84, 68–75.

Putnik, G. D., Putnik, Z., Shah, V., Varela, L., Ferreira, L., Castro, H., Alves, C., & Pinheiro, P. (2021a). Collaborative engineering definition: Distinguishing it from concurrent engineering through the complexity and semiotics lenses. In *Proceedings of the ICAD2021*, Lisbon, 23rd to 25th of June 2021.

Putnik, G. D., Putnik, Z., Shah, V., Varela, L., Ferreira, L., Castro, H., Alves, C., & Pinheiro, P. (2021b). Collaborative engineering: A review of organisational forms for implementation and operation. In *Proceedings of the ICAD2021*, Lisbon, 23rd to 25th of June 2021.

Reardon, C., Tan, H., Kannan, B., & DeRose, L. (2015). Towards safe robot-human collaboration systems using human pose detection. *2015 IEEE International Conference on Technologies for Practical Robot Applications (TePRA)*, 1–6. IEEE.

Robb, D. A., Lopes, J., Padilla, S., Laskov, A., Chiyah Garcia, F. J., Liu, X., . . . Hastie, H. (2019). Exploring interaction with remote autonomous systems using conversational agents. In *Proceedings of the 2019 on Designing Interactive Systems Conference*, 1543–1556. ACM Digital Library, June.

Schuh, G., Potente, T., Wesch-Potente, C., Weber, A. R., & Prote, J-P. (2014). Collaboration mechanisms to increase productivity in the context of industrie 4.0. *Procedia CIRP*, 19, 51–56. ISSN 2212-8271.

Sipsas, K., Alexopoulos, K., Xanthakis, V., & Chryssolouris, G. (2016). Collaborative maintenance in flow-line manufacturing environments: An Industry 4.0 approach. *Procedia CIRP*, 55, 236–241.

Strohkorb, S., Fukuto, E., Warren, N., Taylor, C., Berry, B., & Scassellati, B. (2016). Improving human-human collaboration between children with a social robot. *2016 25th IEEE International Symposium on Robot and Human Interactive Communication (RO-MAN)*, 551–556.

Sturtevant, D. (2018). Modular architectures make you agile in the long run. *IEEE Software*, 35(1), 104–108.

Varela, L., Araújo, A., Ávila, P., Castro, H., & Putnik, G. (2019). Evaluation of the relation between lean manufacturing, Industry 4.0, and sustainability. *Sustainability*, 11(5), 1439. MDPI.

Varela, M. L. R., Putnik, G. D., Manupati, V. K., Rajyalakshmi, G., Trojanowska, J., & Machado, J. (2018). Collaborative manufacturing based on cloud, and on other I4.0 oriented principles and technologies: A systematic literature review and reflections. *Management and Production Engineering Review*, 9(3), 90–99, Polish Academic Sciences, Poland.

Varela, M. L. R., Santos, A. S., & Madureira, A. M. (2014). Collaborative framework for dynamic scheduling supporting in networked manufacturing. *International Journal of Web Portals*, 6(3), 33–51.

Wakefield, K. J. (2014). *How the Internet of Things Is Transforming Manufacturing*. Retrieved July 3, 2018, from http://www.forbes.com/sites/ptc/2014/07/01/how-the-internet-of-things-is-transforming-manufacturing/.

Wang, M., Mei, H., Jiao, W., Jie, J., & Shi, T. (2010). Multi-agent system collaboration based on the relation-web model. *AICI 2010: Artificial Intelligence and Computational Intelligence*, 132–144.

Windeler, A., & Sydow, J. (2001). Project networks and changing industry practices collaborative content production in the German television industry. *Organization Studies,* 22(6), 1035–1060.

Yuan, Y., Li, B., & Kreger, H. (2016). SOA reference architecture: Standards and analysis. In *International Conference on Smart Computing and Communication* (pp. 469–476). Springer, December.

2 Key Enabling Technologies, Methodologies, Frameworks, Tools and Techniques of Smart and Sustainable Systems

Leonel Patrício, Paulo Ávila, Maria Leonilde Rocha Varela, Fernando Romero, Goran D. Putnik, Hélio Castro and Luís Fonseca

CONTENTS

2.1 INTRODUCTION

Industry 4.0 (I4.0) is a German project that combines manufacturing with high-level information technology and digitization (Adolph et al., 2016). It is a revolution in manufacturing and brings innovative perspectives to how manufacturing can participate in new technologies, methodologies, frameworks, tools and techniques—in short, approaches—to achieve maximum production effectiveness and efficiency alongside automation and integration levels, with minimum or optimized use of resources. The effect of this new paradigm derives from the evolution of intelligent or smart factories by refining this concept and further improving and exploring it at higher levels of digitalization and high standard technologies. One central concern and objective of the application of the I4.0 paradigm consists of reaching extremely effective and efficient use of resources, means, products, materials and tools to enable very fast dynamic, flexibility and (re)configurability capabilities to enable customized production and a full integration of all stakeholders and "things" in an organization,

DOI: 10.1201/9781003123866-2

from suppliers to customers and all associated business partners in a large network of partners, which may be organized in different ways, for instance, in virtual organizations or enterprises, distributed or extended manufacturing systems or collaborative networks, to enable more effective and prompt adaptation to the current highly demanding requirements associated with a dynamic and fast-changing globally distributed market, alongside manufacturing and management goals (Kagermann et al., 2013; Deloitte, 2014; Smit et al., 2016; Wittenberg, 2016; Putnik & Ferreira, 2019).

In recent years, intelligent or smart manufacturing has drawn great interest in academia and industry, because it gives competitive advantage to manufacturing organizations, making this type of industry more effective and efficient through the use of advanced information and communication technology (Kagermann et al., 2013; Deloitte, 2014; Smit et al., 2016; Wittenberg, 2016; Putnik & Ferreira, 2019). Moreover, in the current Industry 4.0, companies and underlying manufacturing and management approaches, technologies and systems further need to be sustainable (Varela et al., 2019). In I4.0, the duality of flexibility and productivity is a recurring challenge for organizations that seek to reduce costs and offer a greater range of customized products. The flexibility of manufacturing systems can be understood as the ability to produce a wide variety of products and is considered one of the most important requirements for new applications, for instance, in robotics (Esmaeilian et al. 2016). Moreover, to reach this flexibility, it is further fundamental that available manufacturing systems have a high level of reconfigurability and reliability (Samala et al., 2021a, 2021b; Putnik et al., 2021).

Manufacturers face the impulses of product specification, with the need to increase resource efficiency and reduce product projection times. These stimuli are related to digitization; use of information technologies; and the connection of products, resources and production processes, which are leveraged by the Internet of Things (IoT) (Scheuermann et al., 2015; Rennung et al., 2016). Moreover, the previous requirements should be fulfilled along with economic, social and environmental sustainability needs (Varela et al., 2019), as expressed in Figure 2.1.

The issue of sustainability is becoming very important at the manufacturing level, particularly for industries with intensive practices of resources and energy. Sustainable development is instituting changes in the way manufacturing systems are designed and implemented. Sustainability is emerging as one of the topics in the international governance market, and they are interconnected (Putnik & Ávila, 2016; Almeida et al., 2016).

Researchers have identified the relationships and contributions of Industry 4.0 to sustainability as an emergent and dynamic research subject (Fonseca et al., 2021; Ghobakhloo, 2020). The successful adoption of Industry 4.0 can positively impact sustainability by improving knowledge sharing, collaborative work, production efficiency and productivity (Jena et al., 2020; Machado et al., 2020). Moreover, I4.0 can support novel business models, contribute to cost reduction and customer experience enhancement (Machado et al., 2020; Fonseca et al., 2021) and improve communication and information flows (Linder, 2019). Research results also posit that combining Industry 4.0 technologies with development practices (e.g., Lean) contributes to improved employee morale, reduced lead time, enhanced product quality, customized products and waste reduction (Bogle, 2017; Kamble et al., 2020; Bilge et al.,

2016; Bogue, 2014; Chan et al., 2017). Nevertheless, Industry 4.0 can also have a potential negative influence on sustainability due to cybersecurity risks, labor-saving technologies causing job losses and labor market disruption and increased production and consumption rates leading to resource overconsumption (Beir et al., 2020; Nara et al., 2021).

This chapter identifies the main enabling technologies, approaches, methodologies, methods, techniques, models, tools and platforms for intelligent or smart and sustainable manufacturing systems in Industry 4.0, founded on an organized review of works, including conceptual articles about I4.0 and approaches, technologies and platforms for a sustainable I4.0. The review will be conducted with the following central research question in mind:

Is there increased attention being given to sustainability issues nowadays in the Industry 4.0–oriented smart manufacturing and management context?

To achieve the objective of this work, the rest of this chapter is organized as follows. Section 2.2 presents the research methodology, including the source of information used for conducting the literature search process and subsequent extraction of the most important contributions found in the focused domain of this study. Section 2.3 presents a brief description of the most relevant publications reached through this study and the analysis. Finally, some conclusions and future work are presented in Section 2.4.

2.2 METHODOLOGY

Carrying out a careful literature review is very important to find important insights regarding the state of the art about a specific, and more or less wide, research topic or domain and its evolution in time. In this chapter, a literature review was done to evaluate and analyze existing contributions about sustainable and intelligent manufacturing in companies. Saunders et al. (2016) established an organized review process based on an iterative cycle for defining appropriate keywords for a specific theme by searching important literature and carrying out a corresponding analysis. In this work, a similar methodology was used. In order to deepen the knowledge about the main technologies, methods, techniques, approaches, methodologies, models, tools and platforms for intelligent and sustainable manufacturing systems in Industry 4.0, and based on distributed manufacturing environments or collaborative networks, a study was carried out based on the information in the scientific publications that were searched and analyzed.

Therefore, in this work, several steps were taken, according to the methodology previously described and the main groups of keywords shown in Figure 2.1. According to that methodology, the search, selection and analysis of the various articles directly related to the theme of this literature review are summarized in three main stages: "identifying, evaluating, and synthesizing the existing body of completed and recorded work produced by researchers, scholars, and practitioners" (Fink, 1998). There was a need to follow these steps in order to clarify how this work would be carried out regarding the research theme and further evolve and classify the

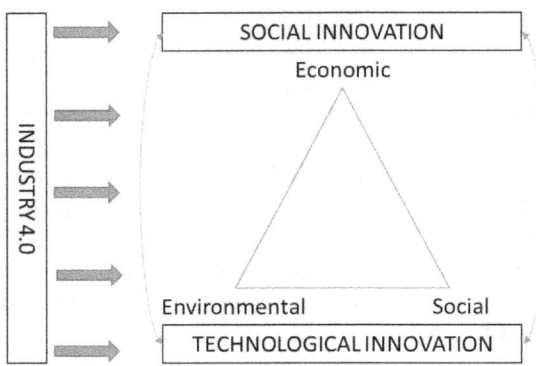

FIGURE 2.1   Sustainable structure applicable to Industry 4.0.

Source: Morrar & Husam, 2017

articles, in addition to the geographic distribution identification, main characteristics and methods, methodologies, technology, approaches, models, techniques, tools and/ or platforms used. For the development of this research work, it was thus necessary to establish the search string and select the academic database to carry out the search process.

The B-ON database was chosen to carry out this research work. This library enables admission to an extensive range of academic publications in international scientific journals and conferences indexed in most well-known indexation systems, for example, Web of Science and Scopus. The B-ON is an extensive database including thousands of peer-reviewed journals and publications in a wide set of fields and arising from different scientific areas. These publications include peer-reviewed articles published before the first trimester of 2021 from several well-known publishers and editors, such as Elsevier, Springer, MDPI and IEEE, among others, that were considered in this study and further analyzed. For this research, a search string was defined by including a set of considered key terms referring to the theme of this work. It is important to note that for each term used, there are several synonyms that can be mentioned and were also used in the search process underlying this work. Table 2.1 shows the search terms and respective synonyms used in the search process.

The search string used in the work for getting articles from the B-ON online library database was based on the key terms and the respective synonyms indicated in Table 2.1:

> **String** = (Group KW1) **AND** (Group KW2) **AND** (Group KW3)
> **String** = *(Approach or technology or method or model or methodology or tool or framework or platform or system or architecture) AND (Smart manufacturing or Industry 4.0, or Industrie 4.0 or I4.0 or Intelligent manufacturing or distributed manufacturing or collaborative networks or collaboration) AND (Sustainability or sustainable or eco-efficient)*

TABLE 2.1

Groups of Key Words Considered for Searching the Literature

Group KW1	Group KW2	Group KW3
(Approach or technology or method or model or methodology or tool or framework or platform or system or architecture)	(Smart manufacturing or Industry 4.0, or Industrie 4.0 or I4.0 or Intelligent manufacturing)	(Sustainability or sustainable or eco-efficient)

TABLE 2.2

Filters Applied to the Search Process

	Articles
Initial result:	717
1 Restrict to: Peer Reviewed	383
2 Type of source: Academic Journals; Conference Materials; Books	382
3 From: 2010 to 2021	380
4 Language: English	372
5 Editor (addleton academic publishers; elsevier b.v.; mdpi ag; elsevier ltd; mdpi; ieee; mdpi publishing; elsevier sci ltd; taylor & francis ltd; elsevier; elsevier science)	334
6 Restrict to: Full Text	249
Final result:	**249**

The string is composed of the main keywords intended to be considered, which are further used to organize the information into three groups (Title, Abstract, Subject Terms). In order to find articles more closely related to the theme of this work, the articles were filtered according to their relevance to the theme. Duplicate articles and those not considered key ones were removed. In addition, parameters that were used for the applied filters were related to articles being peer reviewed and having the full text available. Initially, the research results returned a total of 717 publications. After applying the search filters, the total set of articles decreased to 249, and among these, the ones that were considered more closely or directly related to the research underlying this work were verified. Therefore, the number of articles analyzed dropped to a total of just 20. Table 2.2 shows the filters applied in the search results, as well as the number of articles obtained throughout the search process.

Figure 2.2 represents a flow diagram of the literature search carried out and the respective screening of the topic used in this research work. To analyze the main data of the articles found, it was necessary to carry out two types of characterization. In the first categorization phase, the year of publication and the type of article (research, review, conference, book chapter, journal paper and editorial) were classified. In the second type of analysis, the publications found were characterized as theoretical or conceptual contributions, literature reviews or case studies.

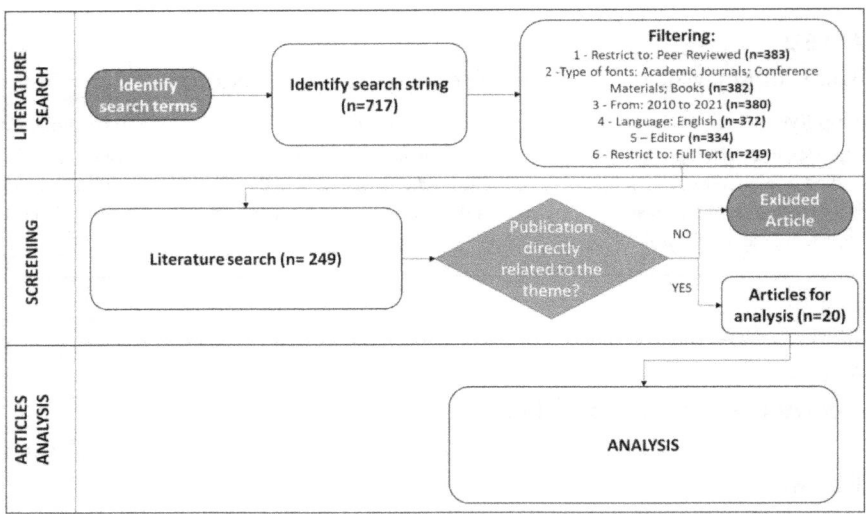

FIGURE 2.2   Flow diagram of literature search and respective screening.

Source: Neves et al., 2020

2.3   ARTICLE SYNTHESIS AND ANALYSIS

In order to deepen the knowledge on enabling technologies, approaches, methodologies, structures, tools and techniques of intelligent and sustainable manufacturing systems for I4.0, a study based on scientific articles was necessary to investigate the comprehensive information on this topic. Throughout this chapter, we will review different contributions from the most relevant papers that address this topic, which falls into the context of I4.0, and the underlying smart manufacturing concept, along with sustainability issues.

Regarding the I4.0, the main pillars that were considered, and for which contributions were found, are related to big data and data analytics, simulation, horizontal and vertical integration, [Industrial] Internet of Things, autonomous robots, the cloud, cyber-physical [production] systems, security, augmented reality and additive manufacturing (Kagermann et al., 2013; Deloitte, 2014; Smit et al., 2016; Putnik & Ferreira, 2019). The collection of articles found and analyzed, as previously mentioned, used the database of the online library B-ON. According to the selection parameters described previously, 19 articles were selected for further analysis, as shown in Table 2.3.

Man and Strandhagen (2017) present an article that discusses possible sustainable business landscapes and proposes a research agenda on how Industry 4.0 can be used to produce sustainable business models, where opportunities for sustainable contributions exist when designing products for longevity. Sustainability means not only being more efficient but also using less raw material and recycling more products. This changes the value proposition, supply chain, customer association and financial validation of a business model. This work addresses the following I4.0 pillars: 3, 5 and 8.

TABLE 2.3

Synthesis of Information Retrieved from Most Relevant Articles Analyzed

Authors	Title	Keywords	Journal
Man and Strandhagen (2017)	An Industry 4.0 Research Agenda for Sustainable Business Models	Sustainability; Business Model; Industry 4.0; Research Agenda	*Procedia CIRP*
Giret et al. (2017)	A Holonic Multi-Agent Methodology to Design Sustainable Intelligent Manufacturing Control Systems	Sustainability; Multi-Agent Systems (MASs); Holonic Control; Manufacturing; Design Method;	*Journal of Cleaner Production*
Yazdi et al. (2018)	An Empirical Investigation of the Relationship between Overall Equipment Efficiency (OEE) and Manufacturing Sustainability in Industry 4.0 with Time Study Approach	Small and Medium Enterprises; OEE; OECD; Manufacturing Sustainability; Time Study; Industry 4.0; Material Handling Systems; Agent-Based Control Architecture	*Sustainability*
Thomas and Kovoor (2018)	Smart Systems Implementation in UK Food Manufacturing Companies: A Sustainability Perspective	Food Manufacturing; Digital Hub; Sustainability Profile; Smart Systems; Survey	*Sustainability*
Varela et al. (2019)	Evaluation of the Relation between Lean Manufacturing, Industry 4.0, and Sustainability	Lean Manufacturing; Industry 4.0; Sustainability; Economic; Environmental; and Social; Structure Equations Modeling	*Sustainability*
Scavarda et al. (2019)	An Analysis of the Corporate Social Responsibility and the Industry 4.0 with Focus on the Youth Generation: A Sustainable Human Resource Management Framework	Sustainable Human Resources; Industry 4.0; Corporate Social Responsibility; Conceptual Framework; Youth Generation	*Sustainability*
Hidayatno et al. (2019)	Industry 4.0 Technology Implementation Impact to Industrial Sustainable Energy in Indonesia: A Model Conceptualization	Industry 4.0; Sustainable Energy; Making Indonesia 4.0; Technology Adoption; Model Conceptualization	*Energy Procedia*
Ren et al. (2019)	A Comprehensive Review of Big Data Analytics Throughout Product Lifecycle to Support Sustainable Smart Manufacturing: A Framework, Challenges and Future Research Directions	Big Data Analytics; Smart Manufacturing; Servitization; Sustainable Production; Conceptual Framework; Product Lifecycle	*Journal of Cleaner Production*

(Continued)

TABLE 2.3 (*Continued*)

Authors	Title	Keywords	Journal
Ghadimi et al. (2019)	Intelligent Sustainable Supplier Selection Using Multi-Agent Technology: Theory and Application for Industry 4.0 Supply Chains	Sustainable Supplier Selection; Industry 4.0; MASs; Cyber-Physical Systems; Industry 4.0 Supply Chain	*Computers & Industrial Engineering*
Lee et al. (2019)	Development of an Intelligent Tool Condition Monitoring System to Identify Manufacturing Tradeoffs and Optimal Machining Conditions	Smart and Sustainable Manufacturing; Artificial Intelligence; Evolutionary Strategies; Tool Condition	*Procedia Manufacturing*
Fatimah et al. (2020)	Industry 4.0 Based Sustainable Circular Economy Approach for Smart Waste Management System to Achieve Sustainable Development Goals: A Case Study of Indonesia	Industry 4.0; Internet of Thing (IoT); Maturity Model; Smart Waste Management; Sustainability; Sustainable Circular Economy; Sustainable Development Goals (SDGs)	*Journal of Cleaner Production*
Bai et al. (2020)	Industry 4.0 Technologies Assessment: A Sustainability Perspective	Industry 4.0; Technology; Sustainability; Hesitant Fuzzy Set; Cumulative Prospect Theory; VIKOR	*International Journal of Production Economics*
Yadav et al. (2020)	A Framework to Achieve Sustainability in Manufacturing Organisations of Developing Economies Using Industry 4.0 Technologies' Enablers	Developing Nations; Empirical Study; Industry 4.0; Manufacturing Supply Chain; New Technologies; Robust Best Worst Method (RBWM); Sustainability	*Computers in Industry*
Villar et al. (2020)	Fostering Economic Growth, Social Inclusion & Sustainability in Industry 4.0: A Systemic Approach	Sustainable Development; Social Inclusion Approach; Soft Systems Methodology; Industry 4.0; SMES Strategy; Manufacturing Sector	*Procedia Manufacturing*
García-Muiña et al. (2020)	Sustainability Transition in Industry 4.0 and Smart Manufacturing with the Triple-Layered Business Model Canvas	Industry 4.0; Sustainability; Manufacturing; Business Model Canvas	*Sustainability*
Ahmad et al. (2021)	Towards Sustainable Textile and Apparel Industry: Exploring the Role of Business Intelligence Systems in the Era of Industry 4.0	Business Intelligence Systems Adoption; Industry 4.0; Sustainability; Textile Industry; Apparel Industry	*Sustainability*

TABLE 2.3

Authors	Title	Keywords	Journal
Yadav et al. (2020)	A Framework to Overcome Sustainable Supply Chain Challenges Through Solution Measures of Industry 4.0 and Circular Economy: An Automotive Case	Sustainable Supply Chain; Challenges; Industry 4.0; Circular Economy; Solution Measures; Best Worst Method; ELECTRE	*Journal of Cleaner Production*
Nara et al. (2021)	Expected Impact of Industry 4.0 Technologies on Sustainable Development: A Study in the Context of Brazil's Plastic Industry	Cloud Computing Search Subject for Cloud Computing, Industry, Internet of Things, Models, Plastics, Robots, Sustainable Development, Brazil	*Sustainable Production and Consumption*
Enyoghasi and Badurdeen (2021)	Industry 4.0 for Sustainable Manufacturing: Opportunities at the Product, Process, and System Levels	Industry 4.0; Sustainable Products; Sustainable Processes; Sustainable Systems	*Resources, Conservation And Recycling Volume*
Costa et al. (2021)	A New Simple, Flexible and Low-Cost Machine Monitoring System	Industry 4.0; Machine Monitoring; Beacon, Bluetooth BLE; Remote Monitoring; Low Cost; Small and Medium Enterprises (SMEs); b-Remote	*Dyna Ingenieria e Industria*

Giret et al. (2017) propose a work focused on a method that helps researchers design sustainable intelligent manufacturing systems, based on holonic technology (Barbosa et al., 2015). The approach centers on identifying the producing elements and hence the style and integration of sustainability-oriented mechanisms within the system specification, providing specific development tools with integrated support for proprietary resources. It is carried out through a set of case study investigations in which the projected technique can be gauged. This work addresses the following I4.0 pillars: 3 and 9.

Yazdi (2018) present a paper with the objective of designing and analyzing the implementation of an intelligent and sustainable material handling system for material distribution using an agent-based algorithm as a control architecture. The study focuses on recognizing and analyzing effective factors in the sustainability of improved processes using a simple model. For this, through expert opinions, effective factors on the sustainability of process improvement activities are determined. This work addresses the following I4.0 pillars: 3 and 9.

Thomas and Kovoor (2018) present an investigation aiming to explore the applicability of intelligent systems in food manufacturing companies in the United Kingdom and to identify the main priority areas and improvement levers for the implementation of such systems. A survey is carried out, including a questionnaire, follow-up interviews and visits to 32 food manufacturing companies in the United Kingdom. The questionnaire and interviews are guided by a unique measurement instrument that the authors developed with a focus on smart systems (SS) technologies. This work presents an original contribution, as it is one of the few academic studies to explore the implementation of SS in industry and provide a new perspective on the main motivators and inhibitors of its implementation. The results suggest that the current turmoil in the sector may be bringing food companies closer to adopting such systems; therefore, this is a good time to define and develop the optimal SS implementation strategy. This work addresses the following I4.0 pillars: 3 and 8.

In Varela et al. (2019), a review of Lean manufacturing (LM), Industry 4.0 (I4.0) and the three pillars of sustainability is put forward, with the main goal of explaining the meaning of these three main subjects. Moreover, the authors focus on a proposed structural equation model based on two exogenous constructs (LM and I4.0) and three endogenous constructs: economic sustainability (EcS), environmental sustainability, and social sustainability (SoS), each construct composed of three manifest variables, with six hypotheses for quantitatively measuring the effects of LM and I4.0 on the three sustainability pillars. Additionally, in order to statistically validate the underlying hypotheses, a collection of 252 valid questionnaires from industrial companies in the Iberian Peninsula (Portugal and Spain) were analyzed. The validation of the proposed model was obtained through the application of corroborative correlational analysis, and further based on the corresponding values of the quality evaluation of the results obtained, based on an intended level of reliability and validity, through corresponding statistical validation measures. In this study, it was possible to realize about the existence of a relation between Industry 4.0 pillars and the sustainability dimensions. The authors believe that these conclusions will positively contribute industrial managers in their daily decision-making in order to promote the transcition of the companies to the digitalization era, while being aware about the importance of aiming at reaching sustainable outcomes.

Hidayatno et al. (2019) present research that aims to discover the systemic impact of the development and implementation of technology from Industry 4.0 for the transition of sustainable energy in developing countries, which eventually needs a valid model conceptualization to act as a standard for future research. The United Nations Industrial Development Organization has defined the relevance of Industry 4.0 and sustainability in global Sustainable Development Goals (SDGs) 7 and 9, stating that digital industrial development will support the growth of the sustainable energy industry. Therefore, the implication will certainly affect all countries with different meanings, including one of the emerging industry countries, Indonesia. In response to this, Indonesia is currently mapping out the way to enter the Industry 4.0 era, making Indonesia 4.0. This work addresses the following I4.0 pillars: 3.

Ren et al. (2019) present a study combining the main technologies of smart manufacturing and the idea of ubiquitous servitization. A comprehensive overview of big data in intelligent manufacturing is undertaken and a conceptual framework proposed from a product lifecycle perspective. As one of the most important technologies for intelligent manufacturing, big data analytics can reveal insights, such as relationships between lifecycle decisions and process parameters, helping industry leaders make more informed business decisions in management environments. This work addresses the following I4.0 pillars: 1, 3 and 9.

Ghadimi et al. (2019) present a study in which they propose a MAS, that is, a system aimed to address the process of evaluating and selecting sustainable suppliers to provide an appropriate communication channel, structured information exchange and visibility between suppliers and manufacturers. In addition, the application of MASs in this process and their natural applicability as one of the technologies that allow the move to the 4.0 supply chain industry are investigated in detail. It turns out that what is proposed in this approach can help decision-makers within manufacturing companies make quick decisions with less human interaction. The merit of the developed MAS is demonstrated through a real-world implementation by a medical device manufacturer. Finally, the limitations and advantages of the proposed approach are presented in conjunction with some observations for future work. Advances in information and communication systems offer immense opportunities for supply chain intelligence and autonomy by establishing stepping stones for Industry 4.0 supply chains (SCs). However, this process has not yet been carried out for the SCs of Industry 4.0, where interconnection, real-time transparency of information, technical assistance and decentralization of members of a physical system (members of the supply chain) are considered the main design principles. This work addresses the following I4.0 pillars: 3 and 9.

Lee et al. (2019) present a tool for condition monitoring of an intelligent system in a research work aimed at identifying manufacturing trade-offs related to sustainability and an ideal set of machining conditions monitoring the status of the machine-tool. In addition, they use a multi-objective optimization based on an evolutionary algorithm to find the ideal operating conditions. Through the increased use of sensors and networked machines in manufacturing operations, artificial intelligence techniques play a fundamental role in deriving significant value from the big data infrastructure. These techniques can inform decision-making and enable the implementation of more sustainable practices in the manufacturing industry. In machining processes, a considerable amount of waste (scrap) is generated as a result of failure to monitor a tool condition. This work addresses the following I4.0 pillars: 1, 2, 3, 5, 8 and 9.

Scavarda et al. (2019) developed research between 1 March 2019 and 2 September 2019 through a bibliographic review involving human resources and deadlines related to the concept of sustainability, Industry 4.0, corporate social responsibility and the young generation. Its public target is the young generation of the world. Two proposals are created after reviewing the literature and collecting data, which allows the elaboration of "an analysis of corporate social responsibility and Industry 4.0 with a focus on the young generation: a sustainable management of human resources structure". The authors contribute with theoretical and practical educational purposes to insert the young citizen into society. This contribution also involves the work of companies in planning and preparing their teams for the development of activities in the communities in their neighborhood, which will allow the creation of new proposals to be presented so that nations can incorporate their young people in the transition labor market and have a sustainable vision for future generations. This work addresses the following I4.0 pillars: 3.

García-Muiña et al. (2020) present a paper with the objective of analyzing the introduction of sustainability into a corporate value proposal through the evolution from a traditional to sustainable business model. Business model innovation is investigated in the case of a producer of ceramic tiles in the district of Sassuolo, Italy. The company has introduced several sustainability practices over the years and, through investments in Industry 4.0 technologies, is able to carry out impact assessments of its production process. The tool applied to the business model transition is the triple-layered business model canvas by Joyce and Paquin. The results illustrate the new company's sustainable value proposal, considering all three pillars of sustainability: environment, economy and society. Despite the limitations resulting from the individual case study, the results can be easily adapted to other ceramic tile companies in the sector. In addition, the authors' research can inspire other manufacturing companies to develop a sustainable business model project. This work explores the still-limited literature on the application of sustainable business model methods in operational scenarios. This work addresses the following I4.0 pillars: 3 and 9.

Ahmad et al. (2021) present a study on one of the determinants of the adoption of business intelligence systems (BIS) with an eye to understanding how BIS can solve sustainability issues in a company with Industry 4.0 technologies. The methodology they use is a qualitative research approach that is applied with 14 semi-structured detailed interviews with 12 of the world's leading Telegraph and Argus (T&A) companies. A snowball and purposeful sampling strategy is used to select participants. The qualitative content analysis technique is used to analyze the interview data. The results revealed several topics, such as sustainability problems in T&A companies, improved value creation processes with leading business intelligence (BI) solutions and difficulties in adopting BIS. Major improvements are seen in apparel retailing because apparel companies are more likely to adopt Industry 4.0 technologies with advanced technologies for business intelligence solutions. The results prove the fundamental role of EcS in the adoption of BIS and Industry 4.0 technologies in T&A companies. This work addresses the following I4.0 pillars: 1, 3 and 8.

Fatimah et al. (2020) present a work whose objectives are to investigate fundamental issues and opportunities and develop a sustainable and intelligent waste management system in Indonesia using technologies from Industry 4.0. The system should provide a multidimensional approach, determine the maturity level of the waste management system via a technical method and meet the objective of designing a new strategy to minimize the problems of waste management. In this work, they present a comprehensive systematic review of the literature; intensive discussions in focus groups and direct observation in Indonesian cities were the approaches used to develop waste management business processes and their system design. The waste business processes consist of mixed collection, classification, transportation, varied treatment and chain disposal. The proposed waste management system project features circular economy processes that can separate municipal waste, identify waste characteristics and determine sustainable waste treatment technologies through the use of the Internet of Things as an integrator. This study contributes to the objectives of the Sustainable Development Goals, such as good health and well-being (SDG 3), drinking water and sanitation (SDG 6), decent work and economic growth (SDG 8), responsible consumption and production (SDG 12) and climate action (SDG 13).

The study proposes a new smart and sustainable waste management project, which can achieve satisfactory economic, social and environmental performance in waste management. This work addresses the following I4.0 pillars: 3 and 4.

Bai et al. (2020) present a structure of measures for sustainability based on the United Nations Sustainable Development Goals incorporating various economic aspects and environmental and social attributes. They also develop a hybrid decision method using multiple situations integrating a hesitant fuzzy set, cumulative prospecting theory and VIKOR. This method can effectively evaluate Industry 4.0 technologies based on their sustainable performance and application. They apply the method using the secondary case information from a report by the World Economic Forum. The results show that mobile technology has the greatest impact on sustainability across all industries, and nanotechnology, mobile technology, simulation and drones have the greatest impact on sustainability in the automotive, electronics, food and beverage and textiles, apparel and footwear industries. The recommendation of this paper is to take advantage of Industry 4.0 adoption technology to improve the impact of sustainability, with the requirement that each technology be carefully assessed in the context of each sustainability dimension. Investment in such technologies should consider the appropriate priorities for investment and promotion. This work addresses the following I4.0 pillars: 2 and 3.

Yadav et al. (2020) present a study with the objective of developing a framework to improve the adoption of sustainability in manufacturing organizations in developing nations using technologies from Industry 4.0. Initially, facilitators who strongly influence the adoption of sustainability are identified through a literature review. In addition, they present large-scale research conducted to allow Industry 4.0–enabling technologies to be included in the structure. Based on empirical analysis, a framework is developed and tested in an Indian manufacturing case organization. Finally, the robust best worst method (RBWM) is used to identify the intensity of influence of each capacitor included in the structure. The results of the study reveal that managerial, economic and environmental facilitators have a strong contribution to the adoption of sustainability. The results of the study will be beneficial for researchers, professionals and policy makers. This work addresses the following I4.0 pillars: 1, 4, 5 and 9.

Villar et al. (2020) present an investigation that proposes a soft systems methodology to deal with the context of sustainable complexity and inclusive industrial development phenomena. Its holistic nature provides useful insights that plan how I4.0 and social inclusion fit into the Mexican context. The theoretical proposal is based on the state of the art of social inclusion in Sector 4.0 and a survey for an I4.0 initiative accessible through a stakeholder system network communication approach. The inclusive strategy is an effort to align root systems for sustainable development with stakeholders for Mexican SMEs in the manufacturing sector. This work addresses the following I4.0 pillars: 3 and 9.

Yadav et al. (2020) present a study that aims to develop a structure to overcome the challenges of sustainable supply chain management (SSCM) through solution measures based on Industry 4.0 and the circular economy. This study identifies a unique set of 28 SSCM challenges and 22 solution measures. In addition, an automotive case organization is used to test the applicability of the framework developed

through the hybrid best worst method (BWM)–elimination and choice express-
ing reality (ELECTRE) approach. Entries for the BWM-ELECTRE approach are
obtained by building a panel of experts within the case organization. Initial entries
are taken for BWM comparisons to calculate the weight of SSCM challenges,
whereas a further comparison of challenges and solution measures is also obtained
using the ELECTRE approach to calculate the final classification of the solution
measures to overcome the SSCM challenges. The results of the case study reveal that
managerial and organizational and economic challenges emerge as the most critical
for the adoption of SSCM. The results of the study will be beneficial for researchers working in the SSCM 4.0 industry and in the domain of the circular economy,
whereas practitioners can use prioritized solution measures to formulate effective
strategies to overcome SSCM adoption failures. This work addresses the following
I4.0 pillars: 3 and 9.

Enyoghasi and Badurdeen (2021) present an investigation with a comparative
analysis examining individual technologies in Industry 4.0 and their potential impact
on sustainable manufacturing. A structure based on clusters of sustainability metrics for products, processes and systems is applied to examine these impacts. The
results reveal that the literature is still limited in identifying opportunities to improve
sustainability at different levels using technologies from Industry 4.0. The impact
on many criteria related to products, processes or sustainability at the system level
due to Industry 4.0 technologies has not yet been examined. Comparative analysis,
and other literature, are used to provide additional guidance for future research and
opportunities on leveraging Industry 4.0 technologies for more sustainable manufacturing. The implications for the industry through providing a framework for identifying potential solutions to improve sustainable manufacturing performance using
Industry 4.0 technologies are also discussed. This work addresses the following I4.0
pillars: 3 and 9.

Nara et al. (2021) develop a study investigating the impacts of Industry 4.0 technologies using the triple-bottom-line perspective for sustainable development. They
present a sustainability-oriented model for assessing the influence of Industry 4.0
technologies on sustainable metrics. The model analyses the impact of Industry 4.0
technologies on several key performance indicators related to sustainable development. The model was tested in the plastics industry, which has a high potential for
technological 4.0 Industry aggregation in emerging economies. A diffuse multi-
criteria TOPSIS method was used to classify Industry 4.0 technologies, identifying
those with the strongest and weakest impacts on sustainable development. As a result, it
was suggested that the Internet of Things, cyber-physical systems, sensors and the
implementation of big data are engines for sustainable development. It is also shown
that these technologies are associated with substantial positive impacts on economic
metrics. However, there was much less positive influence on environmental and
social metrics, suggesting an imbalance in the perspective of the triple bottom line
for the plastics industry. In addition, negative impacts of robots on job creation and
a low influence of cloud computing technologies and systems integration for sustainable development were found. Based on these findings, this work contributes to the
decision-making process by helping managers, process engineers and stakeholders
to understand and estimate the expected impacts of Industry 4.0 technologies on

economic, environmental and social aspects for sustainable development. This work addresses the following I4.0 pillars: 1, 3, 4, 6 and 7.

Costa et al. (2021) propose a system for remote monitoring of equipment in real time that meets the requirements of low cost, simplicity and flexibility. The system monitors the equipment in a simple and agile way, regardless of its sophistication, installation constraints and company resources. A prototype of a system was developed and tested in both laboratory conditions and in a production environment. The proposed architecture of the system comprises a sensor that transmits the machine's signal wirelessly to a gateway, which is responsible for collecting all surrounding signals and sending them to the cloud. During the testing and assessment of the tools, the results validated the developed prototype. As a main result, the proposed solution offers the industrial market a new low-cost monitoring system based in mature and tested technology laid upon flexible and scalable solutions.

The articles previously presented were analyzed based on the nine main pillars of I4.0 (see Table 2.4) and the three pillars of sustainability (see Table 2.5). In both tables, for each of the works, which pillar(s) are addressed was identified, and subsequently the total percentage of the papers that cover each pillar and the percentage of the pillars that are covered by each paper were quantified.

Analyzing the previous tables, it is possible to verify the following:

- The I4.0 pillars most addressed are: horizontal and vertical system integration, additive manufacturing and 3D printing.
- The I4.0 pillars less addressed are: the cloud and cyber security.
- None of the papers covers all the pillars of I4.0.
- The sustainability pillar most addressed is economic, covered by 95% of the papers, and with a large difference compared to the other pillars.
- There are only five papers (25%) that cover all three pillars of sustainability.

2.4 CONCLUSION

This work enabled the identification of the main approaches of intelligent or smart and sustainable manufacturing systems and supply chains or networks in the scope of I4.0 based on a systematic review of the literature. The main objective of this work was to pay particular attention to sustainability issues, as this is considered a central issue of the utmost concern today in a I4.0 context and underlying the smart factory concept. Through this work, it was also possible to verify which of the main I4.0 and sustainability pillars were considered more frequently in research, which are: for I4.0, the integration of horizontal and vertical systems (94%) and additive manufacturing and 3D printing (56%) and for sustainability, the economic dimension (95%), with a large difference compared to the other pillars.

Considering the central research question from Section 2.1, "Is there increased attention being given to sustainability issues nowadays in Industry 4.0?" the results suggest that there is not yet much attention to the pillars of environmental and SoS, which are fundamental dimensions of sustainability. Considering that, there is space to increase research on I4.0 with the aim of achieving a better compromise between industrial development and sustainability.

TABLE 2.4

Pillars of I4.0 Addressed by the Articles Selected in the Research

Research papers	Pillars of I4.0									% Pillars p/article
	Big Data and Data Analytics	Simulation	Horizontal and Vertical Integration	Industrial Internet of Things	Autonomous Robots	The Cloud	Cyber Physical Systems/ Security	Augmented Reality	Additive Manufacturing	
2017 Giret et al.			x						x	22
Man and Strandhagen			x		x			x		33
2018 Yazdi et al.			x						x	22
Thomas et al.			x					x		22
2019 Varela et al.	x	x			x					33
Hidayatno et al.			x							**11**
Ren et al.	x		x						x	33
Ghadimi et al.			x						x	22
Lee et al.	x	x	x		x			x	x	**67**
Scavarda et al.			x							**11**
2020 García-Muiña et al.			x						x	22
Ahmad et al.	x		x					x		33
Fatimah et al.			x	x						22
Bai et al.		x	x							22
Yadava et al.	x		x	x					x	44
Villara et al.			x						x	22
Yadav et al.			x						x	22
2021 Enyoghasi and Badurdeen			x						x	22
Nara, E. et al.	x		x	x		x	x			56
Costa et al.				x						**11**
% Articles p/pillar	28	11	94	22	17	6	6	22	56	

TABLE 2.5

Pillars of Sustainability Addressed by the Articles Selected in the Research

Research Papers	Pillars of Sustainability	Environmental	Social	Economic	% Pillars p/article
2017	Giret, A. et al.			x	33
	Man and Strandhagen	x		x	67
2018	Yazdi et al.			x	33
	Thomas et al.			x	33
2019	Varela et al.	x	x	x	**100**
	Hidayatno et al.	x		x	67
	Ren et al.			x	33
	Ghadimi et al.		x	x	67
	Lee et al.			x	33
	Scavarda et al.		x		33
2020	García-Muiña et al.	x	x	x	**100**
	Ahmad et al.			x	33
	Fatimah et al.	x	x	x	**100**
	Bai et al.	x	x	x	**100**
	Yadava et al.	x		x	67
	Villara et al.		x	x	67
	Yadav et al.			x	33
2021	Enyoghasi and Badurdeen			x	33
	Nara et al.	x	x	x	**100**
	Costa et al.			x	33
% Articles p/pillar		37	37	**95**	

ACKNOWLEDGMENTS

This work has been supported by FCT—Fundação para a Ciência e Tecnologia within the R&D Units Project Scope: UIDB/00319/2020 and EXPL/EME-SIS/1224/2021.

REFERENCES

Adolph, L., et al. (2016) *German Standardization Roadmap: Industry 4.0.* Version2. Berlin: DIN.

Ahmad, M., Muslija, A., Satrovic, E. (2021) Does economic prosperity lead to environmental sustainability in developing economies? Environmental Kuznets curve theory. *Environmental Science and Pollution Research*, 28(18), 22588–22601.

Almeida, A., Bastos, J., Francisco, R., Azevedo, A., Ávila, P. (2016) Sustainability assessment framework for proactive supply chain management. *International Journal of Industrial and Systems Engineering*, 24(2), 198–222.

Bai, C., Dallasega, P., Orzes, G., Sarkis, J. (2020) Industry 4.0 technologies assessment: A sustainability perspective. *International Journal of Production Economics*, 229, 107776.

Barbosa, J., Leitão, P., Adam, E., Trentesaux, D. (2015) Dynamic self-organization in holonic multi-agent manufacturing systems: The ADACOR evolution. *Computers in Industry*, 66, 99–111.

Beier, G., Ullrich, A., Niehoff, S., Reißig, M., Habich, M. (2020) Industry 4.0: How it is defined from a sociotechnical perspective and how much sustainability it includes A literature review. *Journal of Cleaner Production*, 229, 120856.

Bilge, P., Badurdeen, F., Seliger, G., Jawahir, I.S. (2016) A novel manufacturing architecture for sustainable value creation. *CIRP Annals – Manufacturing Technology*, 65, 455–458.

Bogle, I.D.L. (2017) A perspective on smart process manufacturing research challenges for process systems engineers. *Engineering*, 3, 161–165.

Bogue, R. (2014) Sustainable manufacturing: A critical discipline for the twenty-first century. *Assembly Automation*, 34, 117–122.

Chan, F., Li, N., Chung, S.H., Saadat, M. (2017) Management of sustainable manufacturing systems-a review on mathematical problems. *International Journal of Production Research*, 55, 1148–1163.

Costa, J., Ávila, P., Bastos, J., Pinto Ferreira, L. (2021) A new simple, flexible and low-cost machine monitoring system. *DYNA-Ingeniería e Industria*, 96(6).

Deloitte. (2014) *Industry 4.0—Challenges and Solutions for the Digital Transformation and Use of Exponential Technologies*, 1–12. Zurich: Finance, Audit Tax Consulting Corporate, Academia.

Enyoghasi, C., Badurdeen, F. (2021) Industry 4.0 for sustainable manufacturing: Opportunities at the product, process, and system levels. *Resources, Conservation and Recycling*, 166, 105362.

Fatimah, Y.A., Govindan, K., Murniningsih, R., Setiawan, A. (2020) Industry 4.0 based sustainable circular economy approach for smart waste management system to achieve sustainable development goals: A case study of Indonesia. *Journal of Cleaner Production*, 269, 122263.

Fink, A. (1998) *Book Reviews: Conducting Research Literature Reviews: From Paper to the Internet*. London: Sage Publications Inc.

Fonseca, L., Amaral, A., Oliveira, J. (2021) Quality 4.0: The EFQM 2020 model and Industry 4.0 relationships and implications. *Sustainability*, 13, 3107.

García-Muiña, F.E., Medina-Salgado, M.S., Ferrari, A.M., Cucchi, M. (2020) Sustainability transition in industry 4.0 and smart manufacturing with the triple-layered business model canvas. *Sustainability*, 12(6), 2364.

Ghadimi, P., Wang, C., Lim, M.K., Heavey, C. (2019) Intelligent sustainable supplier selection using multi-agent technology: Theory and application for Industry 4.0 supply chains. *Computers & Industrial Engineering*, 127, 588–600.

Giret, A., Trentesaux, D., Salido, M.A., Garcia, E., Adam, E. (2017) A holonic multi-agent methodology to design sustainable intelligent manufacturing control systems. *Journal of Cleaner Production*, 167, 1370–1386.

Hidayatno, A., Destyanto, A.R., Hulu, C.A. (2019) Industry 4.0 technology implementation impact to industrial sustainable energy in Indonesia: A model conceptualization. *Energy Procedia*, 156, 227–233.

Jena, M.C., Mishra, S.K., Moharana, H.S. (2020) Application of Industry 4.0 to enhance sustainable manufacturing. *Environmental Progress and Sustainable Energy*, 39, 13360.

Kagermann, H., Helbig, J., Hellinger, A., Wahlster, W. (2013) Recommendations for implementing the strategic initiative INDUSTRIE 4.0: Securing the future of German manufacturing industry; final report of the Industrie 4.0 Working Group. Forschungsunion.

Kamble, S., Gunasekaran, A., Dhone, N.C. (2020) Industry 4.0 and lean manufacturing practices for sustainable organisational performance in Indian manufacturing companies. *International Journal of Production Research*, 58, 1319–1337.

Lee, W.J., Mendis, G. P., Sutherland, J. W. (2019) Development of an intelligent tool condition monitoring system to identify manufacturing tradeoffs and optimal machining conditions. *Procedia Manufacturing*, 33, 256–263.

Linder, C. (2019) Customer orientation and operations: The role of manufacturing capabilities in small- and medium-sized enterprises. *International Journal of Production Economics*, 216, 105–117.

Machado, C.G., Winroth, M.P., Ribeiro da Silva, E.H.D. (2020) Sustainable manufacturing in Industry 4.0: An emerging research agenda. *International Journal of Production Research*, 58, 1462–1484.

Man, J.C., Strandhagen, J.O. (2017) An Industry 4.0 research agenda for sustainable business models. *Procedia CIRP*, 63, 721–726.

Morrar, R., Arman, H., Mousa, S. (2017) The fourth industrial revolution (Industry 4.0): A social innovation perspective. *Technology Innovation Management Review*, 7(11), 12–20.

Nara, E.O.B., Becker da Costa, M., Baierle, I.C., Schaefer, J.L., Benitez, G.B., Lima do Santos, L.M.A., Benitez, L.B. (2021) Expected impact of Industry 4.0 technologies on sustainable development: A study in the context of Brazil's plastic industry. *Sustainable Production and Consumption,* 25, 102–122.

Neves, A., Godina, R., Azevedo, S., Matias, J. (2020) A comprehensive review of industrial symbiosis. *Journal of Cleaner Production*, 247(20), 119113. DOI: 10.1016/j.jclepro.2019.119113

Putnik, G.D., Ávila, P. (2016) Governance and sustainability (special issue editorial). *International Journal of Industrial and Systems Engineering*, 24(2), 137–143, Inderscience Publishers.

Putnik, G.D., Ferreira, L.G.M. (2019) Industry 4.0: Models, tools and cyber-physical systems for manufacturing (Editorial). *FME Transactions*, 47(4), 659–662.

Putnik, G.D., Pabba, S.K., Manupati, V.K., Varela, M.L.R., Ferreira, F. (2021) Semi-double-loop machine learning based CPS approach for predictive maintenance in manufacturing system based on machine status indications. *CIRP Annals—Manufacturing Technology*, 70(1), 365–368.

Ren, S., Zhang, Y., Liu, Y., Sakao, T., Huisingh, D., Almeida, C.M. (2019) A comprehensive review of big data analytics throughout product lifecycle to support sustainable smart manufacturing: A framework, challenges and future research directions. *Journal of Cleaner Production*, 210, 1343–1365.

Rennung, F., Luminosu, C., Draghici, A. (2016) Service provision in the framework of Industry 4.0. *Procedia—Social and Behavioral Sciences*, 221, 372–377.

Samala, T., Manupati, V.K., Nikhilesh, B.B.S., Varela, M.L.R., Putnik, G.D. (2021a) Job adjustment strategy for predictive maintenance in semi-fully flexible systems based on machine health status. *Sustainability*, 13(9), 5295.

Samala, T., Manupati, V.K., Varela, M.L.R., Putnik, G. (2021b) Investigation of degradation and upgradation models for flexible unit systems: A systematic literature review. *Future Internet, MDPI*, 13(3), 57.

Saunders, M., Lewis, P., Thornhill, A. (2016) *Research Methods for Business Students*. London: Pearson Education Limited.

Scavarda, A., Daú, G., Scavarda, L.F., Goyannes Gusmão Caiado, R. (2019) An analysis of the corporate social responsibility and the Industry 4.0 with focus on the youth generation: A sustainable human resource management framework. *Sustainability*, 11(18), 5130.

Scheuermann, C., Verclas, S., Bruegge, B. (2015) Agile factory—an example of an Industry 4.0 manufacturing process. *2015 IEEE 3rd International Conference on Cyber-Physical Systems, Networks, and Applications* (pp. 43–47). Kowloon, Hong Kong.

Smit, J., Kreutzer, S., Moeller, C., Carlberg, M. (2016) *Industry 4.0*. Elsevier.

Thomas, D., Kovoor, B.C. (2018) A genetic algorithm approach to autonomous smart vehicle parking system. *Procedia Computer Science*, 125, 68–76.

Varela, L., Araújo, A., Ávila, P., Castro, H., Putnik, G. (2019) Evaluation of the relation between lean manufacturing, Industry 4.0, and sustainability. *Sustainability*, 11(5), 1439. DOI: 10.3390/su11051439

Villar, Mendoza-del, Oliva-Lopez, L., Luis-Pineda, E., Benešová, O., Tupa, A., Garza-Reyes, J. A. (2020) Fostering economic growth, social inclusion & sustainability in Industry 4.0: A systemic approach. *Procedia Manufacturing*, 51, 1755–1762.

Wittenberg, C. (2016) Human-CPS interaction—requirements and human-machine interaction methods for the Industry 4.0. *IFAC-Papersonline*, 49–19, 420–425.

Yadav, G., Kumar, A., Luthra, S., Garza-Reyes, J. A., Kumar, V., Batista, L. (2020) A framework to achieve sustainability in manufacturing organisations of developing economies using industry 4.0 technologies' enablers. *Computers in Industry*, 122, 103280.

Yazdi, M. (2018) Risk assessment based on novel intuitionistic fuzzy-hybrid-modified TOPSIS approach. *Safety Science*, 110, 438–448.

3 Real-Time Management-Based Production Scheduling for Sustainability
An Introduction

Cátia Alves and Goran D. Putnik

CONTENTS

3.1 INTRODUCTION

Smart manufacturing has been promoted as a new manufacturing paradigm with the development of advanced technologies, such as the Internet of Things (IoT), artificial intelligence (AI) and cyber-physical systems, among others (Qu, Ming, Liu, Zhang, &

DOI: 10.1201/9781003123866-3

Hou, 2019; Mittal, Khan, Romero, &Wuest, 2019). Research in advanced ICT technologies, under the 4th Industrial Revolution (so-called Industry 4.0), can create a competitive, digital, low-carbon and circular industry (European Commission, 2021) and should be aligned with the Sustainable Development Goals (SDGs) from the United Nations. Industry 4.0 is seen as a powerful instrument for achieving sustainability goals (Varela, Araújo, Ávila, Castro, & Putnik, 2019). Advanced information and communication technology (ICT), in the scenario of Industry 4.0, allows data collection, processing, and decision-making in real time to control (in real time) manufacturing (Alves & Putnik, 2019).

In this context, the creation and implementation of new management paradigms becomes "real", especially in production scheduling, such as real-time management-based production scheduling. In this chapter, the term "real-time management-based production scheduling" will be shortened to "real-time management". Considering smart and sustainable manufacturing, in this chapter, an introduction to real-time management as a new scheduling paradigm is presented, focusing on completion time and cost as two of the principal economic sustainability measures and on CO_2 emissions as one of the principal environmental sustainability measures.

This chapter is further organized as follows. Section 3.2 presents some real-time management definitions and real-time management as a scheduling paradigm. Cyber-physical production systems (CPPSs) are presented in Section 3.3 as an instrument for the effective implementation of real-time management. CPPS instruments could provide effective real-time management, that is, planning and control, of the principal measures of economic and environmental sustainability, which are given in Section 3.4. Section 3.5 presents a model for calculation of these principal measures, including environment dynamics parameters. Further, an example of a real-time management scheduling application under environment dynamics is given in Section 3.6. This chapter ends with Section 3.7, where conclusions are presented.

3.2 REAL-TIME MANAGEMENT

3.2.1 REAL-TIME MANAGEMENT DEFINITIONS

The first known use of "real time" was in 1953, and "real time", "real-time" or "real-time" (noun) is "the actual time during which something takes place" (Merriam-Webster Dictionary, 2014). In literature, some real-time management definitions that are "too broad" can be found, depending on different assumptions and interpretations. Dearden (1966) criticizes Burck's definition, where "a real-time management information system . . . delivers information in time to do something about it", and Martin's definition, where "real-time defined as one that controls an environment by receiving data, processing them and returning results sufficiently quickly to affect the functioning of the environment at that time", as "too broad". According to Dearden (1996), "all management control systems must be real-time systems" under Burck's and Martin's definitions, and "it would be a little silly to plan to provide management with budget performance reports, for instance, if they were received too late for management to take any action". This means that real-time management terms and definitions should be more distinguishable in relation to other traditional management paradigms.

Barlow (2013) defined real time as "squishy" and also criticized the definitions of "real time" as too broad, given that "real time can vary depending on the context in which it is used". Barlow (2013) also presented a discussion concerning issues around real-time management topics such as "how fast is fast" and "how real is real time". Eckerson (2010) also discussed the issue of "how fast is real time", considering that "the definition of real time varies from person to person" and suggesting that the best term is "right time". Houghton, El Sawy, Gray, Donegan and Joshi (2004) presented the definition of "real time" for different contexts: for the factory, it means "as close to real time as possible", and for executive management, it means "once the information has been validated and synchronized among data feeds so that noise has been filtered out".

Real-time management, also called ad hoc management by Borovits and Segev (1977), is related to "maintaining the functioning of the organization" and is "defined as a set of immediate tasks and activities directed towards maintaining the functioning of the organization". Real-time management development "combines working on the company's real opportunities and issues with learning" in Nixon's (1998) "version of management development". Merlo, Vicien and Ducq (2014) refer to real-time management in the design process as unpredictable and needing to be controlled periodically. Considering control, Paiola, Schiavone, Grandinetti and Chen (2021) refer to real-time management as "constant monitoring of . . . complex systems". Other authors refer to control periodicity on a daily basis to "monitor the planned schedules and correct for any disruptions" (Lusby, Larsen, Ehrgott, & Ryan, 2011) or on a real-time basis in "meeting the needs of an individualized customer market" (Tien, Krishnamurthy, & Yasar, 2004). Real-time production control has as its function "to adapt the production system to the changing environment, while preserving efficiency with respect to cost, time and quality requirement" and provide "decisions for specific problems associated with part manufacturing" (Monostori, Kádár, Pfeiffer, & Karnok, 2007), consisting of three key issues: "data acquisition, quick response and instantaneous feedback" (Monostori et al., 2010).

Hwang (2006) uses real-time management in "manufacturing execution systems" in order to "assist management staff of the production line to foresee problems or detect problems as they happen and deal with them immediately to avoid delaying production and affecting output". Real-time management implies "effective, efficient and reactive collaboration among participating entities", once the "parties involved . . . have their own resources and objectives" (Lau, Agussurja, & Thangarajoo, 2008). Although there are different definitions, it could be said that the real-time management discipline is not yet well established, as we can find different interpretations in the literature concerning "What is real-time management?"

3.2.2 Real-Rime Management as a Scheduling Paradigm

This section presents real-time management as a scheduling paradigm. To evaluate the main characteristics of real-time management, it is necessary to present the characteristics of different traditional scheduling paradigms. Therefore, three production scheduling paradigms will be considered (Alves, 2017):

- Fixed horizon and its variant, fixed horizon with reconfiguration
- Rolling horizon
- Real-time management

Fixed-horizon scheduling could be interpreted as scheduling for a specific interval, which is delimited by the allocation of all operation of jobs at the input to machines/resources. This allocation determines the interval of time for processing, that is, the completion time of the given set of jobs (in some other contexts, this could be called time-to-market [TTM]). Due to disturbances and the uncertainty of the environment, some changes in the planned schedule can occur. These changes could be called re-planning, re-scheduling, re-evaluation or even reconfiguration. In the context of this chapter, these terms will be used as synonyms. Considering the fixed horizon, if a reconfiguration occurs, it is necessary to re-plan from the point of reconfiguration to the completion time value, that is, where the allocation of all jobs is concluded. This new schedule, considering reconfiguration, is a variant of fixed-horizon scheduling, and it is called fixed horizon with reconfiguration.

In the rolling horizon scheduling paradigm, the specific interval for scheduling is considered a "rolling period", which is shorter than the interval for the fixed horizon with and without reconfiguration. In this case, when a reconfiguration occurs, it is necessary to reschedule from the point of reconfiguration until the next "rolling period". These traditional paradigms are widely used in scheduling. Fixed horizon could be more suitable for static environments, while real-time management is more suitable for dynamic environments. Due to the environment dynamics intrinsically present in the new era of Industry 4.0, the real-time management scheduling paradigm emerged. In real-time management, only scheduling between two reconfigurations is considered. At the limit, the scheduling (planning) period should tend to zero, that is, as short as possible. In accordance with the concept of "real time", real-time management scheduling implies the capacity for making a schedule at each instant of time. Consequently, it implies that the planning horizon is, at the limit, equal to one time unit.

Even if the planning horizon is equal to one time unit, there is always a plan between two reconfigurations. This means that real-time management scheduling could be considered a sequence of fixed horizon scheduling. So, these different scheduling paradigms could be classified as a function of the length and/or type of planning horizon and the existence of reconfigurations. Thus, the planning horizon could be interpreted as the time for execution of scheduling for a specific interval that is delimited by the projected completion time, for example, delimited by the time required to complete all tasks of all jobs scheduled (Alves & Putnik, 2019).

3.3 CYBER-PHYSICAL PRODUCTION SYSTEMS AS AN INSTRUMENT FOR REAL-TIME MANAGEMENT

The real-time management paradigm becomes real and applicable because of emergent advanced ICT technologies in the Industry 4.0 scenario. In manufacturing systems, some ICT technologies, such as ubiquitous computational systems, permit control and operation of production systems anywhere and anytime, for which real-time management is inherent (anytime from anywhere) (Putnik & Wang, 2017). Putnik (2012) presented an architecture for advanced manufacturing and enterprises, such as ubiquitous and cloud manufacturing systems, with embedded real-time management capacity. Further, Ferreira et al. (2013) proposed a cloudlet architecture for "dashboard to monitor a Cloud Ubiquitous Manufacturing System", with integrated pragmatics-oriented services (Ferreira et al., 2014), as an instrument for effective real-time management.

Another instrument for effective implementation of real-time management is cyber-physical systems (CPSs), or, more exactly, cyber-physical production systems in manufacturing systems. The definition of a cyber-physical production system is similar to Lee's (2008) definition of a cyber-physical system as "integrations of computation with physical processes. Embedded computers and networks monitor and control the physical processes, usually with feedback loops where physical processes affect computations and vice versa". G. Putnik, Ferreira, Lopes and Z. Putnik (2019) presented a spectrum for cyber-physical system definitions and models and introduced an intelligent and effective cyber-physical system concept called I-CPS.

Cyber-physical production systems enhance actual manufacturing system control and management processes and, consequently, the decision-making process (Alves & Putnik, 2019), providing adaptability, reconfigurability and fast responsiveness (Jian, Jin, Mingcheng, & Li, 2017) in real time for effective real-time management. Thus, cyber-physical production systems have different capabilities over traditional manufacturing systems, especially in the capability of real-time decision-making. Through the implementation of a large number of sensors within the physical manufacturing system, making it an object of the Internet of Things and smart, it is possible to capture different sensor data for controlling functionalities and sensor data for the principal measures of economic and environmental sustainability calculation. The sensor data captured is an input for simulations and decision-making processes, whose output will affect the physical manufacturing system. Simulations run on the CPPS's "digital layer" (Alves & Putnik, 2019), and these simulations could include environment dynamics parameters. For an environment dynamics description, see Alves (2017).

3.4 PRINCIPAL REAL-TIME MANAGEMENT MEASURES FOR ECONOMIC AND ENVIRONMENTAL SUSTAINABILITY

As presented in the previous section, through the implementation of a large number of diverse sensors, it is possible to capture, in real time, sensor data from the physical manufacturing system for economic and environmental sustainability performance measure calculation. It is important to note that the data for economic and environmental sustainability measures could also be collected manually, but considering the real-time management paradigm, and to provide full real-time effectiveness, data from sensors within the CPPS is obtained "instantaneously". Some principal measures for economic and environmental sustainability can be found in the literature on scheduling, such as real-time management and rolling horizon (considered by some authors a real-time management model), for example:

- time (see, for example: Clark and Clark, 2000; Mohammadi, Ghomi, Karimi, and Torabi, 2010; Nossal and Galla, 1997; Yin, Zou, and Zou, 2011)
- cost (see, for example: Prasad and Chetty, 2001; Clark and Clark, 2000; Li and Ierapetritou, 2010)
- energy saving and consumption reduction (see, for example: Yin et al., 2011) that correspond to economic and environmental sustainability effects
- CO_2 emissions: this is another interpretation of the criteria energy saving and consumption reduction (EC, 2010)

These measures can be analyzed independently or combined, for example, in Manupati, Bose, Agrawal, Putnik and Varela (2019), which addressed the flexible job shop scheduling problem considering energy consumption, completion time and processing cost measures for sustainable manufacturing. Considering cost measures with real-time monitoring through visual feedback, some enterprises improved their efficiency from 44% to 87% in three months concerning production gains, others immediately improved by 29% in production efficiency and other maximized their earnings from $9,000 to $30,000 per month (Vorne, 2014). A 3D visualization management system at a company studied by Qian, Tu and Lou (2019) improved information transparency in the production process, reduced production costs and raised production efficiency more than 50%. Concerning energy saving, with a real-time energy management initiative, 13.2% of energy savings was achieved (Henderson, Waltner, & Council, 2013). Concerning CO_2 emissions, this criterion is strongly promoted by the European Commission, which requires the drastic reduction of CO_2 emissions, which real-time management for manufacturing should follow for production scheduling and energy consumption management.

3.5 A MODEL FOR CALCULATION OF PRINCIPAL ECONOMIC AND ENVIRONMENTAL SUSTAINABILITY MEASURES

The model presents the calculation of real-time management measures for economic and environmental sustainability, namely time (defined as completion time), cost and CO_2 emissions. The model presented focuses on the fixed horizon scheduling paradigm, abstracting previous and subsequent schedulings (between reconfigurations) in which the initial state (from the schedule) is equal to the state of the previous schedule at the moment of starting the rescheduling or reconfiguration (see Section 3.2.1) (Alves, 2017). This is because fixed-horizon scheduling is considered the "elementary building block" of which scheduling paradigm models are constructed (see Alves, 2017; Putnik et al., 2017). The environment dynamics model is presented in Alves (2017), and it will be considered in the model presentation.

3.5.1 Model Assumptions and Notations

The model has the following assumptions:

- Each job corresponds to one process plan. A job is composed of a sequence of tasks, where each task corresponds to one operation described in the process plan. Only one task/operation of the same job can be processed at a time.
- No-preemption is assumed. Each operation, once started, must be completed before another operation may start on that resource.
- One task/operation can be executed at only one resource.
- No cancellation is assumed, and each job must be completed.
- Setup and transportation times are integrated into the task/operation processing time (in the context of this chapter).
- Time consists of discrete periods: 1,2, . . .
- No relation exists between jobs; that is, jobs are independent.
- Resources are available at t_0 and can process only one operation at a time.

The notations of the model are presented in the following:

i	Index for the number of the job
j	Index for the number of operations
k	Index for resources
n	Number of jobs to be processed
m	Total number of machines
J_i	Job i
O_{ij}^k	Operation j of job i processed in resource k
n_i	Total number of operations for job i
μ_{ij}	Set of candidate resources for operation j of job i
M_k	Resource k
p_{ij}^k	Processing time for operation j of job i processed in resource k
c_{ij}^k	Cost of processing operation j of job i processed in resource k
e_{ij}^k	CO_2 emissions for operation j of job i processed in resource k
$t(\tau)$	Time as a function of time in the dynamic environment model
τ	Time in the dynamic environment model
R_τ	Reconfiguration at time t
W_{ij}	Waiting time between operations of J_i
r_i	Release time of J_i
d_i	Due date of J_i
$s_{i,j}$	Start time of operation j of job i

(Continued)

(Continued)

et_{ij}	End time of operation j of job i
$tconf$	Configuration time
$trconf$	Reconfiguration time
wp	Weighted value for processing time
wc	Weighted value for cost
we	Weighted value for CO_2 emissions
a_k	Next available time of resource k
C	Total completion time
C_i	Completion time of J_i
$Cost$	Cost of processing J_i
$CO2$	CO_2 emissions of processing J_i

3.5.2 MODEL FOR COMPLETION TIME CALCULATION

The model for completion time calculation is defined as follows. Each operation O_{ij} $(j = 1, \ldots, n_i)$ can be processed in a set of alternative resources (machines, enterprises, etc.), where k $(k = 1, \ldots, m)$, for the operation O_{ij}^k, represents the resource that can process the jth operation of J_i. By the kth resource, it is not meant that the jth is processed on M_k (although it may coincide). So, there is a set of candidate resources $\mu_{ij} \in \{M_1; \ldots; M_m\}$ that can process O_{ij}. O_{ij} may be processed in only one of the candidate resources in μ_{ij}. Associated with each operation O_{ij}^k $(k \in \mu_{ij})$ is the processing time (p_{ij}^k), the cost of processing (c_{ij}^k) and the CO_2 emission value (e_{ij}^k).

The values of processing time, cost and CO_2 emissions can be considered static or dynamic values. The dynamic value is considered in accordance with the environment model (see Alves, 2017 for more details), as at the moment of reconfiguration, the time values may change. R_τ represents the reconfiguration at time t. So, the values of processing time, cost and CO_2 emissions are a function of time and of the reconfiguration time, $p_{ij}^k(t(\tau), R_\tau)$, $c_{ij}^k(t(\tau), R_\tau)$ and $e_{ij}^k(t(\tau), R_\tau)$, respectively. τ represents time in the dynamic environment model. The release time r_i of J_i is considered the effective start time of O_{i1}. The due date d_i of J_i is assumed to be the end time of O_{i,n_i}. In the context of this chapter, there is a correspondence between d_i and the completion time C_i. W_{ij} is the waiting time between operations of J_i. A Gantt diagram is presented in Figure 3.1 to illustrate these definitions and notations, considering two different jobs: J_a and J_b.

The calculation of the completion time C (in some other contexts, it could be called time-to-market) corresponds to the maximum C_i for all jobs:

$$C = \text{Max}\{C_i \mid i = 1, \ldots, n\}$$

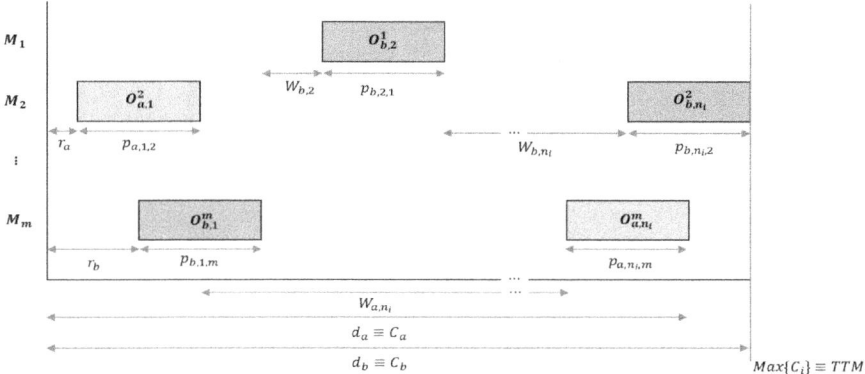

FIGURE 3.1   Gantt diagram for the definitions and notation representation for J_a and J_b.

The completion time of J_i is the time when J_i is finished:

$$C_i = tconf + r_i + \sum_{j=1}^{n_i} \left(W_{ij} + p_{ij}^k \left(t(\tau), R_\tau \right) \right), \forall k$$

where

- the waiting time W_{ij} is calculated by the difference between the start time of O_{ij}, $s_{i,j}$, and the end time, et_{ij-1}, of the $O_{i,j-1}$.

$$W'_{ij} = s_{i,j} - et_{ij-1}, j \neq 1$$

$$W_{ij} = \begin{cases} W'_{ij} & if\ W'_{ij} > treconf \\ treconf, & otherwise \end{cases}$$

- the technological constraint where the operation $j+1$ for the job J_i should start after the operation j is finished is given by:

$$s_{i,j+1} > et_{ij}$$

- k is the selected resource, defined by:

$$k = \max \left(wp * \left(p_{ij} \left(t(\tau), R_\tau \right) + a^* \right)' + wc * c'_{ij} \left(t(\tau), R_\tau \right) + we * e'_{ij} \left(t(\tau), R_\tau \right) \right)_{\mu_{ij}}, \mu_{ij} \in \{ M_1; \ldots; M_m \}$$

where wp, wc and we are the weighted values for the processing time, cost and CO_2 emissions value, respectively, and a_k^* is the updated next available time of the resource k. So, if O_{ij} is processed on the candidate resource k from $\mu_{ij} \in \{ M_1; \ldots; M_m \}$, p_{ij} is the time of processing O_{ij} on M_k, c_{ij} is the cost of processing O_{ij} on M_k and e_{ij} is the CO_2 emission value of processing O_{ij} on M_k. All the values of $\left(p_{ij}^k + a_k^* \right)'$, $c_{ij}^{k'}$

and $e_{ij}^{k\prime}$ are normalized by linear interpolation. For each possible machine, the time, cost and CO_2 emissions are calculated. For the maximum value of time, cost and CO_2, the value of 1 is attributed, and for the minimum values, the value of 10 is used. The other values, which are not the maximum and the minimum, are normalized through linear interpolation.

a_k^* is calculated as the maximum between the next available time of the resource k (a_k) and the end time of the previous operation $O_{i,j-1}$ for all jobs (et_{ij-1}^k).

$$a_k^* = \max\left\{a_k, et_{ij-1}^k\right\}, \forall k$$

If the weighted value result is the same for two or more resource candidates, the resource k selected is the one that has less next available time; that is, if $a_k > a_{k+1}$, select $k+1$; otherwise, if $a_k < a_{k+1}$, select k. If the next available time is $a_k = a_{k+1}$ for the candidate resources, the first resource, k, will be selected.

3.5.3 COST AND CO_2 EMISSIONS CALCULATION

The model for cost and CO_2 emissions calculation, as important performance measures for economic and environmental sustainability, is given in the following. The total cost, $Cost$, is calculated by the sum for costs, c_{ij}^k, for all jobs i and all operations j processed by the selected resource k:

$$Cost = \sum_{i=1}^{n}\sum_{j=1}^{n_i} c_{ij}^k \left(t(\tau), R_\tau\right), \forall k$$

Similarly, the total CO_2 emission calculation, $CO2$, is given by the sum for CO_2 emission, e_{ij}^k, for all jobs i and all operations j processed by the selected resource k:

$$CO2 = \sum_{i=1}^{n}\sum_{j=1}^{n_i} e_{ij}^k \left(t(\tau), R_\tau\right), \forall k$$

3.6 AN EXAMPLE APPLICATION OF REAL-TIME MANAGEMENT SCHEDULING UNDER ENVIRONMENT DYNAMICS

This example considers environment dynamics, which is discussed in Alves (2017), and this example only focuses on the real-time management paradigm.

3.6.1 INITIAL CONDITIONS

Consider two jobs with three operations each:[1]

$$J_1 = \left\{O_{11}, O_{12}, O_{13}\right\} \text{ and } J_2 = \left\{O_{21}, O_{22}, O_{33}\right\}$$

Each J_i is independent and cannot be related. The O_{ij} for each job has its own technological precedence; for example, O_{12} cannot start before O_{11} ends.

For each operation O_{ij}, consider the following set of candidate resources:[2]

$$O_{11} = \{M_1; M_2\}, O_{12} = \{M_1; M_2\}, O_{13} = \{M_2\}$$

$$O_{21} = \{M_1; M_2\}, O_{22} = \{M_1; M_2\}, O_{23} = \{M_2\}$$

Considering the dynamics environment, at the actual moment (R_0), there are forecasts for the O_{ij} processing time, cost and CO_2 emissions in all candidate resources. However, in this example, the values of c_{ij}^k and e_{ij}^k will be excluded, as this calculation is not critical for this illustrative example.

3.6.2 EXAMPLE

There are several heuristic methods, called dispatching rules, to define the jobs that will be dispatched by each order. In this example, the following rule will be used: the order of jobs at the input is created randomly and maintained along the time, and the operations for all jobs are launched by the first operation that can be executed independently, satisfying the technological precedence of operations for each job. For this case, the following order for dispatching is considered: O_{11}, O_{21}, O_{12}, O_{22}, O_{13}, O_{23}. In this example, the weighted values are: $wp = 1$, $wc = 0$ and $we = 0$. This means that we will only consider the processing time for resource selection k:

$$k = \max\left(p_{ij}\left(t(\tau), R_\tau\right) + a^*\right)_{\mu_{ij}}, \mu_{ij} \in \{M_1; \ldots; M_m\}$$

TABLE 3.1
Extracted Processing Times for O_{ij}^k Considered for This Example at R_0

R_0		M_1		M_2		
		OP_1	OP_2	OP_1	OP_2	OP_3
t_0	J_1	3	9	2	10	4
	J_2	8	6	2	4	5
t_1	J_1	1	1	4	4	2
	J_2	6	3	7	5	1
t_2	J_1	7	7	5	4	3
	J_2	6	10	7	2	10

(Continued)

TABLE 3.1   (*Continued*)

R_0		M_1		M_2		
		OP_1	OP_2	OP_1	OP_2	OP_3
...	J_1	...	...	...	...	...
	J_2	...	...	...	...	...
t_8	J_1	2	9	10	1	1
	J_2	3	9	2	8	5
...	J_1	...	...	...	...	...
	J_2	...	...	...	...	...
t_{12}	J_1	4	6	1	7	2
	J_2	9	4	4	5	1

Table 3.1 presents an extract of the processing times for O_{ij}^k considered for this example at R_0, for the maximum time horizon of 12 time units, which corresponds to the interval between reconfigurations.

In real-time management scheduling, the schedule is realized for all operations O_{ij} that can start before and within the reconfiguration period. In this example, two reconfigurations on $t_{12}{}^3$ and t_{24} are considered. So, the first schedule is made for the period $[t_0, t_{12}]$ and the second for $[t_{12}, t_{24}]$. By dispatching order, the first operation to be allocated is O_{11}. This operation has two candidate resources, M_1 and M_2. The values of the processing time of O_{11} (p_{11}^k) on each candidate resource correspond to the time in Table 3.1 of $p_{11}^k(t(0), R_0)$. Considering $a_k = 0$, this means that all machines are available at moment t_0. The summary of resource selection is the following:

O_{11}		M_1	M_2
a_k	(1)	0	0
$et_{i,(j-1)}^k, j \neq 1$	(2)	-	-
a_k^*	(3)	0	0
p_{11}^k	(4)	3	2
Total	(3)+(4)	3	**2**

FIGURE 3.2   Gantt diagram for J1OP1 allocation.

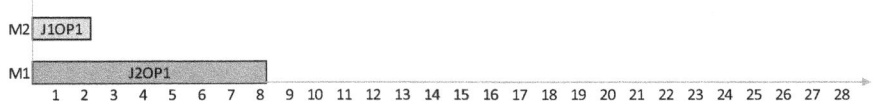

FIGURE 3.3 Gantt diagram for J2OP1 allocation.

The resource selected to process O_{11} is M_2, with a processing time of 2 time units (Figure 3.2). So a_2 should be updated to 2. As M_1 is still not occupied, a_1 will be kept at 0.

The next O_{ij} to be launched is O_{21}. Similarly to O_{11}, the summary of the resource selection is the following:

O_{21}		M_1	M_2
a_k	(1)	0	2
$et_{i,(j-1)}^k, j \neq 1$	(2)	-	-
a_k^*	(3)	0	2
p_{21}^k	(4)	8	7
Total	(3)+(4)	**8**	9

For the case of $a_2^* = 2$, p_{21}^2 corresponds to the value in the pair $\left(t(2), R_0 \right)$. The resource that will process O_{21} is M_1, with a processing time of 8 time units (Figure 3.3).

Following the dispatching order, the next operation to be processed is O_{12}. As operation O_{11} will end at t_2, the resource selection is the following:

O_{12}		M_1	M_2
a_k	(1)	8	2
$et_{i,(j-1)}^k, j \neq 1$	(2)	2	2
a_k^*	(3)	8	2
p_{12}^k	(4)	9	4
Total	(3)+(4)	17	6

The resource that will process O_{12} is M_2, with a processing time of 4 time units (Figure 3.4).

Following the dispatching order, the next operation to be processed is O_{22}. As operation O_{21} will end at t_8, and considering:

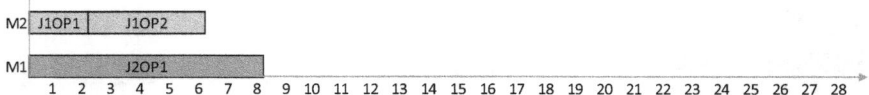

FIGURE 3.4   Gantt diagram for J1OP2 allocation.

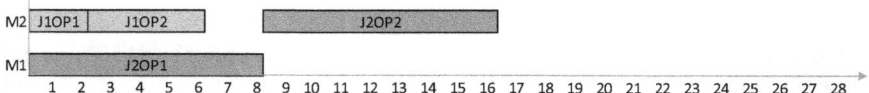

FIGURE 3.5   Gantt diagram for J2OP2 allocation.

$$a_k^* = \max\left\{a_k, et_{ij-1}^k\right\}$$

then,

$$a_2^* = \max\left\{a_2, et_{21}^2\right\} = \max\left\{6,8\right\} = 8$$

O_{22}		M_1	M_2
a_k	(1)	8	6
$et_{i,(j-1)}^k, j \neq 1$	(2)	8	8
a_k^*	(3)	8	8
p_{22}^k	(4)	9	8
Total	(3)+(4)	17	**16**

The resource that will process O_{22} is M_2, with a processing time of 3 time units (Figure 3.5).

So, the first schedule comprises the dispatching of the following operations: O_{11}, O_{21}, O_{12} and O_{22}, as the remaining operations, O_{13}, O_{23}, can only start after the reconfiguration period (Figure 3.6).

At the reconfiguration moment, it is necessary to verify the concluded operations, the operations in execution and the operations not started (Figure 3.7 and Table 3.2).

At the moment of the reconfiguration at t_{12}, new forecasts are considered, with new values, in R_{12}. These values may be different from the values at R_0 (Table 3.1) because of the environment's dynamics. Let's consider the following values at R_{12} (Table 3.3).

As the operations O_{13} and O_{23} could start before t_{24}, they can be scheduled within the reconfiguration period $[t_{12}, t_{24}]$. O_{13} has only one candidate resource, so that is the one that will be selected (Figure 3.8).

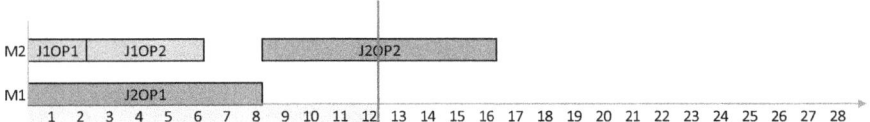

FIGURE 3.6 Gantt diagram for the first period $[t_0, t_{12}]$.

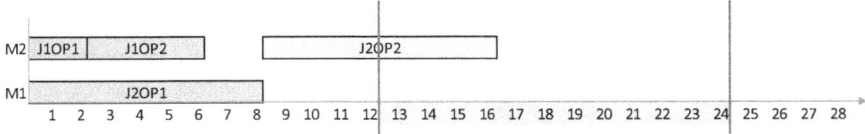

FIGURE 3.7 Gantt diagram with reconfiguration at t_{12} for identification of the operations in execution and the operations which are not started.

TABLE 3.2
Operations Status at R_{12}

Concluded	Processing	Not Started
O_{11}	O_{22}	O_{13}
O_{21}		O_{23}
O_{12}		

TABLE 3.3
Processing Times for O_{ij}^k Considered for This Example at R_{12}

R_{12}		M_1		M_2		
		OP_1	OP_2	OP_1	OP_2	OP_3
t_{12}	J_1	6	8	2	1	7
	J_2	5	4	5	9	4
...	J_1	...	...	...	...	...
	J_2	...	...	...	...	...
t_{16}	J_1	6	10	6	1	5
	J_2	8	3	10	6	3
...	J_1	...	...	...	...	...
	J_2	...	...	...	...	...
t_{21}	J_1	4	5	8	1	8
	J_2	1	3	10	8	5

(Continued)

TABLE 3.3   *(Continued)*

R_{12}		M_1		M_2		
		OP_1	OP_2	OP_1	OP_2	OP_3
...	J_1	...	...	...	...	...
	J_2	...	...	...	...	...
t_{24}	J_1	4	3	4	5	4
	J_2	2	7	4	8	3
	J_2	2	3	10	7	8

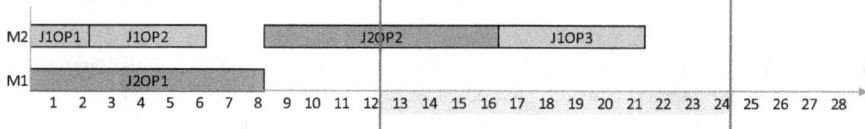

FIGURE 3.8   Gantt diagram for J1OP3 allocation.

O_{13}		M_2
a_k	(1)	16
$et_{i,(j-1)}^k, j \neq 1$	(2)	6
a_k^*	(3)	16
p_{13}^k	(4)	5
Total	(3)+(4)	**21**

In this case, the value of p_{13}^2 is different (although in general, it could be equal), as it is considered the value contained in (t_{16}, R_{12}) and not (t_{16}, R_0).

The procedure is the same for O_{23} (Figure 3.9):

O_{23}		M_2
a_k	(1)	21
$et_{i,(j-1)}^k, j \neq 1$	(2)	16
a_k^*	(3)	21
p_{23}^k	(4)	5
Total	(3)+(4)	**26**

At the reconfiguration moment, t_{24}, there are no operations to be launched (Figure 3.10).

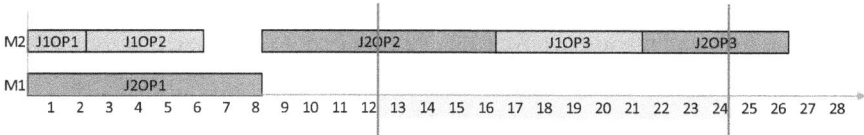

FIGURE 3.9   Gantt diagram for J2OP3 allocation considering t_{12}.

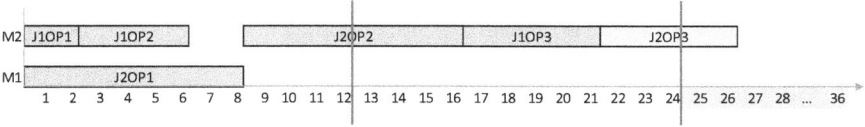

FIGURE 3.10   Gantt diagram with reconfiguration at t_{24} for identification of the operations in execution and the operations that are not started.

As there are no more operations to launch, it is possible to calculate the completion time (or, in other contexts, the TTM). The completion time in this case is 26:

$$C_1 = 0 + (0+2) + (1+3) + (10+5) = 21$$
$$C_2 = 0 + (0+8) + (0+8) + (5+5) = 26$$
$$C = \text{Max}\{C_1, C_2\} = \text{Max}\{21, 26\} = 26$$

3.7   CONCLUSIONS

This chapter presented an introduction to real-time management as a scheduling paradigm. Due to the research on smart manufacturing, real-time management has become "real" and its implementation effective. Cyber-physical production systems are an instrument for effective real-time management implementation, which could allow decision-making in real time based on simulation results from the sensor data collected on sustainability performance measures as a factor of sustainable manufacturing. The example presented in this chapter is very simple and does not reflect a true real-time management scenario in terms of real case. This example has a primarily didactic purpose for perception, adoption and implementation of real-time management principles. Real cases actually involve dozens or hundreds of jobs and machines organized in large networks, cumulative delay as one of the principal factors that influence (re)scheduling and other parameters and factors. Real case simulations are presented in the literature and were the basis for this chapter, in particular Alves (2017) and Alves and Putnik (2019).

Future research directions could include other sustainability measures, also including social sustainability as one of the measures in real-time management under the sustainability triple bottom line. To ensure total implementation and effective management of real-time management, especially considering CPPSs, research should consider exploitation of advanced and emergent software and hardware development under Industry 4.0.

ACKNOWLEDGMENTS

This work has been supported by FCT—Fundação para a Ciência e Tecnologia within the R&D Units Project Scope: UIDB/00319/2020.

NOTES

1 Please note that in this illustrative example, it was considered that jobs have the same number of operations, but in general, jobs can contain different numbers of operations.
2 Please note that in this practical example, it was considered that for all jobs, the candidate resources for O_{ij} are the same, but different O_{ij} could have different candidate resources.
3 Due to the length limit of this chapter, in this illustrative example, the reconfiguration period between two reconfigurations was considered 12 units of time. For a true real-time management example, the period between two reconfigurations should be 1 unit of time.

REFERENCES

Alves, C. (2017). *Modelling and evaluation of "fixed horizon", "rolling horizon" and "real time management" production scheduling paradigms in ubiquitous production networks under conditions of dynamic environments for economic and environmental sustainability*. University of Minho.

Alves, C., & Putnik, G. (2019). Cyber-physical production system (CPPS) decision making duration time impact on manufacturing system performance. *FME Transactions, 47*(4), 675–682.

Barlow, M. (2013). *Real-time big data analytics: Emerging architecture*. O'Reilly Media, Inc.

Borovits, I., & Segev, E. (1977). Real-time management—an analogy. *Academy of Management Review, 2*(2), 311–316.

Clark, A. R., & Clark, S. J. (2000). Rolling-horizon lot-sizing when set-up times are sequence-dependent. *International Journal of Production Research, 38*(10), 2287–2307.

Dearden, J. (1966). Myth of real-time management information. *Harvard Business Review, 44*(3), 123–132.

EC (2010). *EUROPE 2020: A European strategy for smart, sustainable and inclusive growth*. Brussels: COM(2010) 2020 of 3.3.2010.

Eckerson, W. W. (2010). *Performance dashboards: Measuring, monitoring, and managing your business*. John Wiley & Sons.

European Commission (2021). *Strategic plan 2021–2024*. European Union.

Ferreira, L., Putnik, G., Cruz-Cunha, M. M., Putnik, Z., Castro, H., Alves, C., & Shah, V. (2014). Dashboard services for pragmatics-based interoperability in cloud and ubiquitous manufacturing. *International Journal of Web Portals (IJWP), 6*(1), 35–49.

Ferreira, L., Putnik, G., Cruz-Cunha, M., Putnik, Z., Castro, H., Alves, C., & Varela, M. L. R. (2013). Cloudlet architecture for dashboard in cloud and ubiquitous manufacturing. *Procedia CIRP, 12*, 366–371.

Henderson, P., Waltner, M., & Council, N. R. D. (2013). *Real-time energy management: A case study of three large commercial buildings in Washington, D.C.* NRDC Case Study, CS:13–07-A.

Houghton, R., El Sawy, O. A., Gray, P., Donegan, C., & Joshi, A. (2004). Vigilant information systems for managing enterprises in dynamic supply chains: Real-time dashboards at Western Digital. *MIS Quarterly Executive, 3*(1), 19–35.

Hwang, Y.-D. (2006). the practices of integrating manufacturing execution systems and Six Sigma methodology. *The International Journal of Advanced Manufacturing Technology, 31*(1–2), 145–154.

Jiang, Z., Jin, Y., Mingcheng, E., & Li, Q. (2017). Distributed dynamic scheduling for cyber-physical production systems based on a multi-agent system. *IEEE Access, 6*, 1855–1869.

Lau, H. C., Agussurja, L., & Thangarajoo, R. (2008). Real-time supply chain control via multi-agent adjustable autonomy. *Computers & Operations Research, 35*(11), 3452–3464.

Lee, E. A. (2008). Cyber physical systems: Design challenges. In *2008 11th IEEE international symposium on object and component-oriented real-time distributed computing (ISORC)* (pp. 363–369). IEEE.

Li, Z., & Ierapetritou, M. G. (2010). Rolling horizon based planning and scheduling integration with production capacity consideration. *Chemical Engineering Science, 65*(22), 5887–5900.

Lusby, R. M., Larsen, J., Ehrgott, M., & Ryan, D. (2011). Railway track allocation: Models and methods. *OR spectrum, 33*(4), 843–883.

Manupati, V. K., Bose, S. C., Agrawal, R., Putnik, G. D., & Varela, M. L. R. (2019). Integration of process planning and scheduling in an energy-efficient flexible job shop: A hybrid moth flame evolutionary algorithm. In *Integration of Process Planning and Scheduling* (pp. 115–134). CRC Press.

Merlo, C., Vicien, G., & Ducq, Y. (2014). Interoperability modelling methodology for product design organisations. *International Journal of Production Research, 52*(15), 4379–4395.

Merriam-Webster Dictionary. (2014). *First known use of real time*. United States, Society of North America.

Mittal, S., Khan, M. A., Romero, D., & Wuest, T. (2019). Smart manufacturing: Characteristics, technologies and enabling factors. *Proceedings of the Institution of Mechanical Engineers, Part B: Journal of Engineering Manufacture, 233*(5), 1342–1361.

Mohammadi, M., Ghomi, S. F., Karimi, B., & Torabi, S. A. (2010). Rolling-horizon and fix-and-relax heuristics for the multi-product multi-level capacitated lotsizing problem with sequence-dependent setups. *Journal of Intelligent Manufacturing, 21*(4), 501–510.

Monostori, L., Csáji, B. C., Kádár, B., Pfeiffer, A., Ilie-Zudor, E., Kemény, Z., & Szathmári, M. (2010). Towards adaptive and digital manufacturing. *Annual Reviews in Control, 34*(1), 118–128.

Monostori, L., Kádár, B., Pfeiffer, A., & Karnok, D. (2007). Solution approaches to real-time control of customized mass production. *CIRP Annals-Manufacturing Technology, 56*(1), 431–434.

Nixon, B. (1998). Creating the futures we desire—getting the whole system into the room: Part I. *Industrial and Commercial Training, 30*(1), 4–11.

Nossal, R., & Galla, T. M. (1997). Solving NP-complete problems in real-time system design by multichromosome genetic algorithms. In *Proceedings of the SIGPLAN 1997 workshop on languages, compilers, and tools for real-time systems*. University Park, Pennsylvania, United States, CiteSeerX.

Paiola, M., Schiavone, F., Grandinetti, R., & Chen, J. (2021). Digital servitization and sustainability through networking: Some evidences from IoT-based business models. *Journal of Business Research, 132*, 507–516.

Prasad, P., & Chetty, O. K. (2001). Multilevel lot sizing with a genetic algorithm under fixed and rolling horizons. *The International Journal of Advanced Manufacturing Technology, 18*(7), 520–527.

Putnik, G. (2012). Advanced manufacturing systems and enterprises: Cloud and ubiquitous manufacturing and an architecture. *Journal of Applied Engineering Science, 10*(3), 127–134.

Putnik, G., Alves, C., Ávila, P., Ferreira, L., Lopes, N., & Shah, V. (2017). Fixed horizon scheduling as an elementary building block for the fixed horizon with reconfiguration, rolling horizon and real time management production scheduling paradigms. *Proceedings of 2100 Projects Association Joint Conferences, 5*, 36–42.

Putnik, G., Ferreira, L., Lopes, N., & Putnik, Z. (2019). What is a cyber-physical system: Definitions and models spectrum. *FME Transactions, 47*(4), 663–674.

Putnik, G. D., & Wang, L. (2017). Ubiquitous and cloud enterprise for manufacturing. *International Journal of Computer Integrated Manufacturing, 30*(4–5), 344–346. doi: 10.1080/0951192X.2017.1292698

Qian, X., Tu, J., & Lou, P. (2019). A general architecture of a 3D visualization system for shop floor management. *Journal of Intelligent Manufacturing, 30*(4), 1531–1545.

Qu, Y. J., Ming, X. G., Liu, Z. W., Zhang, X. Y., & Hou, Z. T. (2019). Smart manufacturing systems: State of the art and future trends. *The International Journal of Advanced Manufacturing Technology, 103*(9), 3751–3768.

Tien, J. M., Krishnamurthy, A., & Yasar, A. (2004). Towards real-time customized management of supply and demand chains. *Journal of Systems Science and Systems Engineering, 13*(3), 257–278.

Varela, L., Araújo, A., Ávila, P., Castro, H., & Putnik, G. (2019). Evaluation of the relation between lean manufacturing, Industry 4.0, and sustainability. *Sustainability, 11*(5), 1439.

Vorne (2014). Now loading a short message from Vorne. *Frontiers in Physiology*, 61.

Yin, E., Zou, F., & Zou, F. (2011). Research on the optimization method of virtual enterprise's task scheduling problems. *Modern Applied Science, 5*(1), 68.

4 Human–Robot Collaboration in Industry
Threats and Opportunities

Dario Antonelli and Giulia Bruno

CONTENTS

4.1 INTRODUCTION

The interest and even the enthusiasm that worldwide industrial operators and academic researchers have shown toward human–robot collaboration (HRC) in the framework of Industry 4.0 is indubitable and surely positive. Several European countries have introduced supporting policies that made the transformation of existing equipment with new collaborative models that are compliant with Industry 4.0 guidelines affordable (Almada-Lobo, 2016). Principal robot manufacturers already produce collaborative robot models expressly designed to allow human workers to share workspaces with robots. For the introduction of collaborative robots in industry, the starting point was the ISO Technical Standard 15066, published in 2016, that specifies the safety requirements to be followed by collaborative industrial robots. ISO TS 15066 completed the revised safety standards for industrial robots, ISO 10218-1 and ISO 10218-2, published in 2011 (Rosenstrauch, 2017). Since 2016, collaborative robotics was made possible in industry with all the necessary security guarantees.

DOI: 10.1201/9781003123866-4

From the viewpoint of robot manufacturers, collaborative robots became a new market opportunity for their products, extending the existing one. Until then, industrial robots were only used in fully automated work environments. The high fixed costs of an automatic plant and the need for complete standardization of the process limited the use of robots to factories where mass production was carried out. Following Hägele (2016): "Today's industrial robots are mainly the result of the requirements of capital-intensive large-volume manufacturing, mainly defined by the automotive, electronics, and electrical goods industries". Today, industrial production is evolving towards smaller batches and customized products. It needs a degree of flexibility that it is not obtainable in a fully automated workstation. The collaborative human–robot workcell is a solution to this issue.

The technical challenges for robot manufacturers are described by De Santis (2008) with the term *dependability*. Dependability is the sum of different, partly correlated attributes:

- Safety: avoid damaging users and the working environment.
- Availability: service readiness.
- Reliability: continuity during the time of available service.
- Integrity: absence of incorrect changes to the system.
- Maintainability: ability to overcome equipment outages.

These challenges have been addressed by robot manufacturers, and it is possible to assert that the current generation of collaborative robots guarantees dependability to a degree sufficient for many industrial applications by conforming to the safety standards in ISO/TS 10218. From the viewpoint of industrial robot users, there is an appealing prospect: being able to execute processes for small-volume productions, joining the accuracy and the force of a robot with the dexterity and the flexibility of a human. The most frequent objections are not technical but related to social and psychological factors: the fear that robots could steal jobs from humans and the lack of confidence when working in a team with a non-human partner. Therefore, human–robot collaboration is drawing agreement but also some concerns regarding psychological stress induced in the human team. Although these issues have been widely investigated either by economists or psychologists (see Hinds, 2004), they are not discussed in the present chapter. It is reasonable to expect that the massive spread of collaborative robots in every production activity will significantly reduce the fears that many workers have of them due to lack of familiarity with the new machines.

The real dimensions of the problems and issues raised by HRC extend more widely than the ethical and economic problems and involve different engineering fields. It is important to remark that the industrial robot is only one of the many components in a modern production system. The basic working unit in a modern factory, before and after Industry 4.0, is the workcell.

A workcell is an arrangement of resources in a manufacturing environment to improve the quality, speed and cost of the process. Workcells are designed to improve these by improving process flow and are based on the principles of Lean Manufacturing.

(Womack, 1991)

TABLE 4.1

Aspects of Human–Robot Collaborative Workcell

Workcell Design	Process Planning	Robot Programming	Process Execution
Model the collaborative workstation	Work breakdown structure	Task learning	Collaborative interaction, H-R
Model the manufacturing/ assembly process	Task assignment	Trajectory generation	Program execution
Model the tools	Task scheduling	Virtual/augmented reality (VR/AR) feedback	Collision detection

Robotic workcells are therefore composed of robots, their end-effectors, associated sensors and workpiece feeding and positioning devices. It is an expensive industrial system, whose design poses several problems that involve expertise in manufacturing, production planning and logistics. In order to convert a manual workcell to a collaborative one, it is necessary to review and redesign all its components and even its organization.

It is helpful to describe with Table 4.1, on different abstraction levels, the problems that are encountered during the design of a collaborative human–robot workcell.

Therefore, the design of a collaborative workcell should consider different abstraction levels and the different types of tasks that constitute the industrial production process. First, the cell must be designed by considering the different requirements of humans and robots, reflected in a classic layout through the differences between manual and automatic working cells. To execute a collaborative process, it is necessary to define a complete and comprehensive model of the whole workcell, considering the robots; the rotating table, if any; the part to be produced; the tools to be used; the human workers to put beside the robots; and the kind of skills provided by different agents. The concepts behind such a workcell cannot be borrowed either from manual or automatic production. The layout is different, the tools are different and the skills of human workers should be complementary to the skills of the robot. In other words, to design a collaborative cell, experience in fully automated factories is of little help.

A second abstraction level refers to process planning. In manual processes, often all workers are able to execute every operation, and tasks are assigned in such a way as to have balanced workloads. Differently from manual or automatic processes, in HRC, it is crucial to specify which tasks must be performed collaboratively and which separately by humans and by robots. Additionally, collaboration is a general concept that can have many different applications: sharing workspace at different times, working simultaneously on different areas of the workspace, robot assistance and hand guiding (ISO-TS 15066). The correct preparation of activities should start from the definition of the tasks, sub-tasks and sequences. Then the tasks should be assigned to the human, to the robot or to both jointly. Task sequence should be scheduled, and capacity planning on parallel tasks should be performed in order to avoid overload of operators.

The third level in the collaborative process design is the interactive programming of the robot. Standard robot programming, trajectory-oriented using the teach pendant or offline robot simulation, is a time-consuming process that requires a specialized robot programmer. New programming approaches were introduced to exploit the collaborative skills of robots: training by demonstration, manual guidance and voice or gesture task-oriented commands. These new programming techniques have been tested in research environments, but their maturity level is far from acceptable to permit their diffusion in the factory. Eventually, during task execution, a number of activities are performed and need to be properly prepared. They are enabled by the collaborative interaction that passes through a human–machine interface. With the present state of technology, the interface and the terms used to interact with robots cannot be expressed in natural language. They must be application based and defined using a domain-specific language. Another activity is the execution of the program and the error recovery plan. When working with humans, it is not possible to avoid many small deviations from the programmed and expected behavior. As an example, the human can end his task before or after the allotted time, can change the assembly order and can produce many other small variations that require the capacity to reprogram the robot. In automatic production, if there is an even small nonconformity, robots stop and control personnel intervenes to reset the standard work conditions. Working with humans, this would lead to continuous halts. Therefore, a wider range of flexibility is required during program execution on the side of the robot. Safety management and collision detection are additional activities that raise some concerns. The new ISO 15066 standard applies different prescriptions depending on the collaboration type.

Based on the scheme of Table 4.1, this chapter is structured as follows. Section 4.2 describes the state of the art of human–robot collaboration by comparing HRC with corresponding automatic and manual production. Section 4.3 introduces operational procedures applicable to both manual and automatic workstations. The basic concepts of the assembly process are represented by an ontology. This ontology is related to a second one describing the relations between the elements of a robotic cell. The two ontologies are coupled and instantiated, allowing redesign of the robotic workcell with HRC. Section 4.4 describes the problem of collaborative task assignment. Section 4.5 stresses the importance of feedback in the communication between human and robot. Section 4.6 discusses the importance of flexibility to ensure the success of a collaborative operation by applying continuous adaptations to the working program during task execution.

4.2 COLLABORATIVE WORKCELL AS A THIRD WAY BETWEEN FULLY AUTOMATIC AND FULLY MANUAL

Despite its only recent industrial application, HRC has been the subject of research for 30 years. At the beginning, the research started by designing human-compatible robotic hardware (Kazerooin, 1990), then expanded to control modalities that were human friendly (Luh & Hu, 1998). Further research topics in HRC were social characteristics of human–robot interaction (Breazeal, 1998), the development of natural user interfaces (Kang, 1995) and the definition and description of collaborative tasks

(Klingspor, 1997). Goodrich's survey (2007) reviews the progress in HRC up to the 2000s and offers a comprehensive description of possible interactions with robots, how they could be exploited and the main difficulties to overcome. Nevertheless, industrial robots are neglected, as every kind of interaction at automatic workstations was completely absent.

The evolution of the research considered not only physical interaction but also cognitive interaction. Communication systems between human and robot range from a brain–computer interface (Bell, 2008) to AR (Mavrikios, 2013) or advanced speech recognition (Mavridis, 2015). Progress is helped by the advances in machine learning (ML) and dedicated robotic hardware (Kim, 2013). Robotic devices that are HRC oriented are tentatively classified in ISO/TS 15066. ML is now the core of knowledge representation in robotics (Nguyen, 2011). Despite various applications in the field of autonomous, bio-inspired, health or service robots, industrial exploitation of HRC has been limited to the elimination of safety fences (Helms, 2002). After Industry 4.0, there has been an increase in the number of research projects focused on industrial applications (Almada-Lobo, 2016).

The characteristics that distinguish an HRC workcell have been classified by Wang (2017) as:

- *"intuitive and multimodal programming environment:* the human worker doesn't need an in-depth knowledge of the workcell",
- *"zero-programming:* workers communicate with robots via gestures, voice commands, manual guidance and other forms of natural inputs without the need of coding",
- *"immersive collaboration:* with the help of different devices, e.g. screens, goggles, wearable displays", the robot can communicate and collaborate with the worker, and
- *"Context/situation dependency:* the system should be capable of interleaving autonomous human with robot decisions based on inputs from on-site sensors and monitors".

It is worth noting the focus on programming time and straightforwardness. A big issue in having humans and robots working together in a factory layout is in the longer learning time of the robot compared with the human. To understand the perspectives and threats in HRC, it is useful to compare conventional automated processes and manual ones. In automatic processes, programming time is roughly related to the number of significant waypoints along the work trajectory. Applying this concept, it is possible to assert that programming time depends on the complexity and the length of the trajectories executed in the task. Another way to see it is that tasks requiring dexterous movements require longer programming time. On the contrary, humans, having better dexterity, don't need a longer time to execute dexterous tasks.

For the sake of discussion, it is worth classifying collaborative work into three classes: fully automated, with only robots assigned to all tasks; fully manual, without robots and with only human workers; and collaborative, where humans and robots divide up tasks. The three scenarios are described in terms of process time by Figure 4.1.

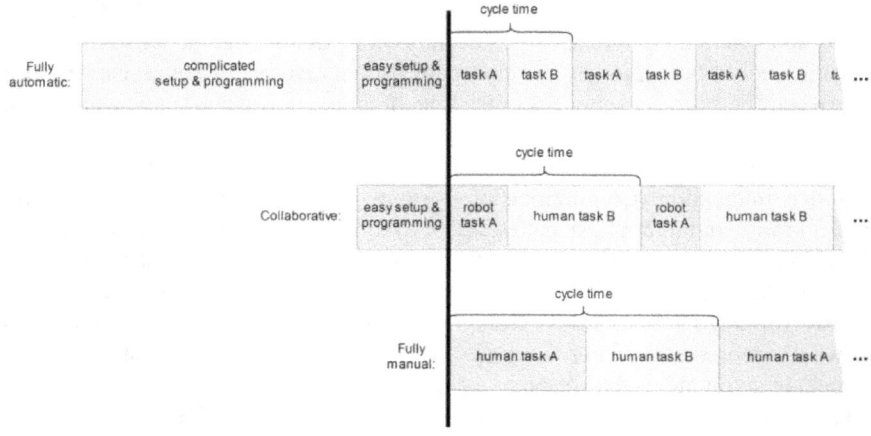

FIGURE 4.1 Cycle time in the three considered scenarios: fully automated, collaborative and fully manual.

It is apparent that, according to the proposed example, robots require shorter cycle time with respect to humans in the execution of both simple or dexterous tasks. Also, humans require the same time to execute dexterous or simple tasks, even if it is much longer than robots. Conversely, automatic production adds long programming time to the cycle time. How much this affects total process time is a function of the number of repeated tasks. In the collaborative scenario, robot tasks are executed faster than human tasks. The programming time is short, as tasks requiring dexterity have been assigned to humans. Summing up the considerations: in large batches, a fully automated scenario should be preferred, as the time spent in robot programming is a fraction of execution time; in very small batches, the manual scenario is advantageous, as no time is spent in robot programming. There is an exploitation area between large and small batches where HRC could bring benefits in terms of less programming time and less execution time. The strengths of both human and robot are combined.

The choice of the preferred work organization is therefore dependent on production volumes, programming time and cycle time. In the considered example, two welding tasks are executed by a human or robot. The first is welding along a line, while the second follows an Archimedean spiral that is a complex trajectory. Figure 4.2 shows the total execution times as a function of the number of consecutive welds executed. There are three curves, for the automatic, manual and collaborative process. In the collaborative case, the robot executes the linear welding and the human executes the spiral welding. The experiment has been executed in a laboratory, and welding was only simulated (the welding gun was turned off).

Obviously, if the programming time of an easy task is shorter or if the difference in time between human and robot tasks is smaller, the area where HRC is competitive can increase. Robot programming for easy tasks can be shortened by preprogramming some repetitive operations; as an example, palletizing several parts could be programmed for actual robots by just giving the pallet dimensions, the position to

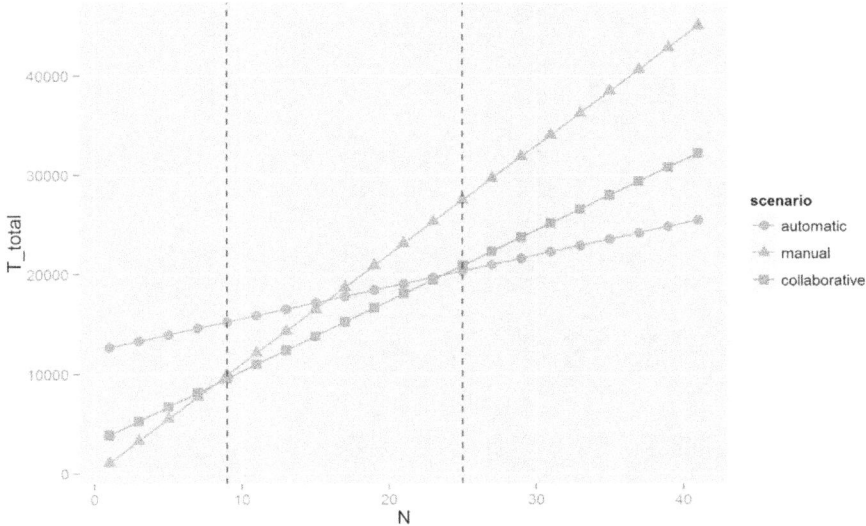

FIGURE 4.2 Comparison of the total times in the three scenarios showing the range of the batch size in which the collaborative scenario is the most efficient. $K = 10$.

place the first part on the pallet and the number of parts in the rows and columns. Another way to reduce programming time is to program the robot through manual guidance or even to employ programming-by-demonstration techniques (Argall, 2009).

4.3 MODELING A COLLABORATIVE WORKCELL

The design of an HRC system starts from a shared knowledge of both workcell components and working activities. The harmonization of terminology related to collaborative robotics is a relevant task to exchange information among researchers in the field and to define a standard that can be shared by users and producers (Haidegger et al., 2013). To this end, ontologies were defined to provide a knowledge base for human–robot collaborative systems.

An ontology is a formal structure used to describe relevant concepts and relations between them in a given domain (Guarino, 1998). Ontologies are used in a variety of different contexts to explicit the knowledge of interest in each specific domain.

In the robotic context, the Ontologies for Robotics and Automation Working Group (ORA WG) defines a core ontology for robotics and automation (CORA), which provides concepts common to industrial robots (Prestes et al., 2013). CORA is an extension of SUMO, Suggested Upper Merged Ontology. SUMO is an open-source upper ontology used in several domains (Niles & Pease, 2001). CORA focuses on defining robots, along with any related object. CORAX is an ontology that stands between CORA and SUMO. CORAX represents concepts and relations of common subdomains that are too general to be included in CORA (Fiorini et al., 2015).

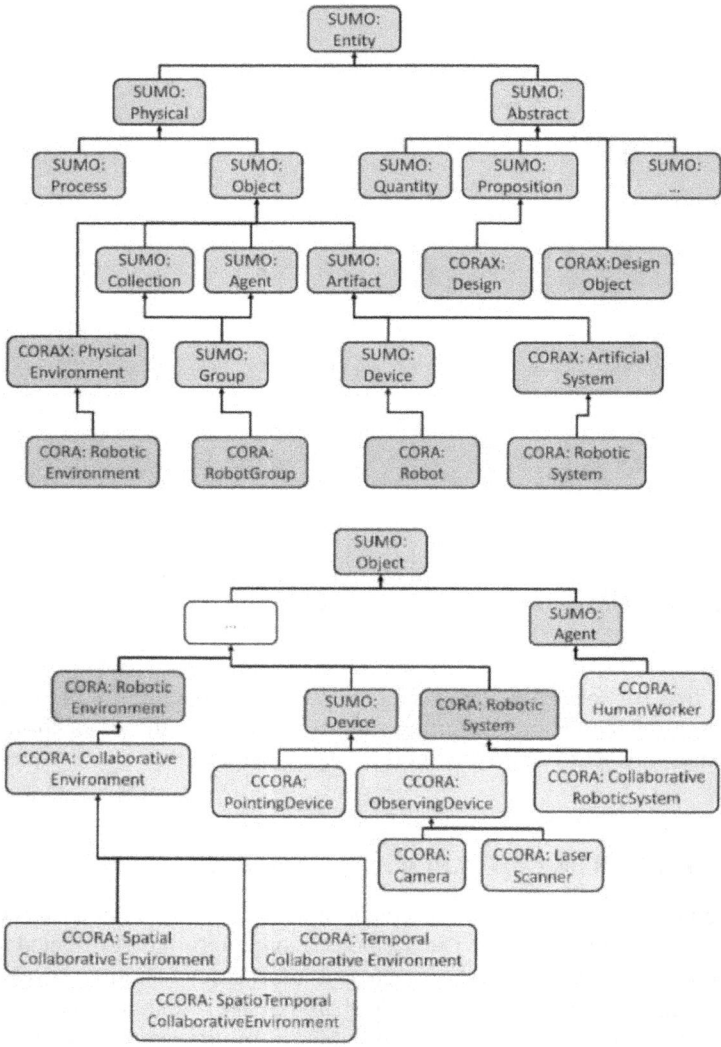

FIGURE 4.3   Left: CORA concepts related to SUMO and CORAX. Right: CCORA concepts related to CORA and SUMO.

Figure 4.3 (left) shows the main concepts of CORA related to SUMO and CORAX. Entity is the more inclusive category, given as a separate partition of physical and abstract entities. Physical entities are those extending in space or time; abstract entities are those that do not allow a description in terms of spatial or temporal dimensions. Descending from Entity, there are further specializations: Physics consists of Object and Process, Abstract consists of Quantity, Attribute, SetOrClass, Relation and Proposition. Presently SUMO has more than 500 entities. The CORAX ontology defines concepts that are too general to be present in CORA, essential for

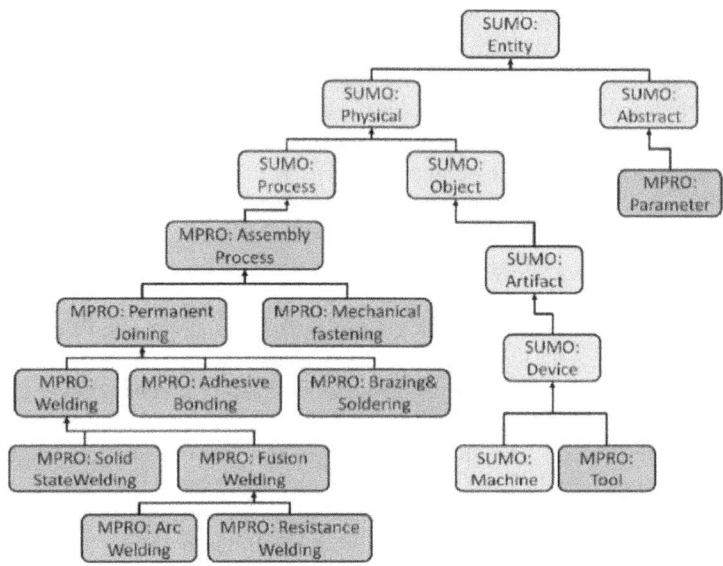

FIGURE 4.4   Merging MPRO and SUMO ontologies.

modeling but not explicitly or completely representable in SUMO. Examples of entities included in the CORAX ontology are physical environment, interaction, artificial system, processing device, robot movement, human robot communication and
machine–machine communication. CORA focuses on defining a robot, along with
specifying other related entities. The CORA ontology is made of the following concepts: Robot part, Robot interface, Robot group, Robotic system (further divided into
Single Robotic System and Collective Robotic System) and Robotic environment.

In order to adapt the ontology to HRC systems, Antonelli and Bruno (2017) add
collaborative concepts to CORA ontology, generating a new extended ontology called
Collaborative CORA (CCORA). Figure 4.3 (right) shows the main concepts of CCORA
with reference to SUMO and CORA. CCORA introduces the definition of collaborative
robotic system as an entity made of robots, human operators and dedicated devices. HRC
requires two specific families of devices for observing the work area and pointing to
relevant points in the work area. Furthermore, CCORA defines the collaborative robotic
environment as an environment in which only collaborative robots operate. A collaborative robotic environment admits three specializations: spatial collaboration environment,
temporal collaboration environment and speed- and force-limited collaboration environment. The requirements in terms of devices depend on the environment specialization.
When humans and robots share the workspace but work in different time instants, the
spatial collaboration environment, there is the need for a monitoring device that detects
the presence and movement of humans. For this application, a laser scanner is fit. A laser
scanner gives a two-dimensional map of the robot's distance from every moving point.
The information is sufficient to avoid the human when the robot is moving.

Regarding working operations, Antonelli and Bruno (2017) defined the manufacturing process ontology (MPRO) covering the knowledge associated with

manufacturing operations. Figure 4.4 shows the components of the MPRO ontology associated with the SUMO ontology (Antonelli et al., 2014).

MPRO ontology has the main classes Machine, Tool, Manufacturing process and Process Parameter (Bruno, 2015). Tool is the component that is used during a process or for directly working on the workpiece. As an example, the Welding tool is applied by different welding processes comprising the Welding mold, Solid state tool and Fusion welding tool. The Fusion welding tool is primarily referred to the Electrodes, which can be Consumables or Non-consumables. Among the Consumables electrodes, there are coated electrodes, Electrogas welding electrodes, Flux cored electrodes, Wire electrodes and GMAW guns.

Conversely, the manufacturing process generates two sub-classes: Process operation and Assembly operation. Assembly operation generates two sub-classes: Permanent joining process and Temporary joining process. Permanent joining process can be subdivided into Welding, Brazing, Soldering or Adhesive bonding. Welding process is the parent class of Fusion welding and Solid state welding. Fusion welding is parent to Arc Welding, Resistance Welding, Oxyfuel gas welding and others. The Process Parameter class contains the parameters of manufacturing processes that need to be explicitly set up, such as cutting speed of the drill process, welding temperature and cooling water pressure. With these ontologies at one's disposal, it is possible to populate them by instantiating different production processes and different workcell components. The ontologies are the basis to define the design requirements of the HRC work environment.

4.4   TASK DEFINITION, ASSIGNMENT AND SCHEDULING

Production planning is commonly written as an optimization problem with the minimization of time and costs of production as its objective function and, as constraint functions, the capacity of workstations and the balancing of the workload. Common approaches at production control try to decompose the full factory optimization problem into a set of subproblems, that is, the optimization of individual workcells by finding local optimal solutions. If the different cells are sufficiently uncorrelated, the superposition of local optimal solutions approximates the global optimum well. When production is completely or partly manual, an additional management problem is workload balancing among the workers, provided that labor is flexible so that every worker can execute every task. This is hardly the case in a collaborative work cell with robots and humans.

As humans and robots have different skills that should be exploited as much as possible, HRC requires a different strategy for work assignment by exploiting individual task execution skills and without any balance of workload between humans and robots. Figure 4.5 presents a skill-exploiting strategy that assign tasks to humans and robots based on task indicators. The sequence of tasks involved in the process is the initial step of the procedure.

The next step is the evaluation of selected collaborative indicators for every task. Indicators make reference to the skills humans or robots have, like dexterity, accuracy and so on. Indicators don't require the exact evaluation of a performance index but can be expressed by using categorial values, or even logical values.

The purpose is not the assessment of tasks on the basis of how fit they are for a robot or human. They are just used to feed a classifier that assigns the tasks to a

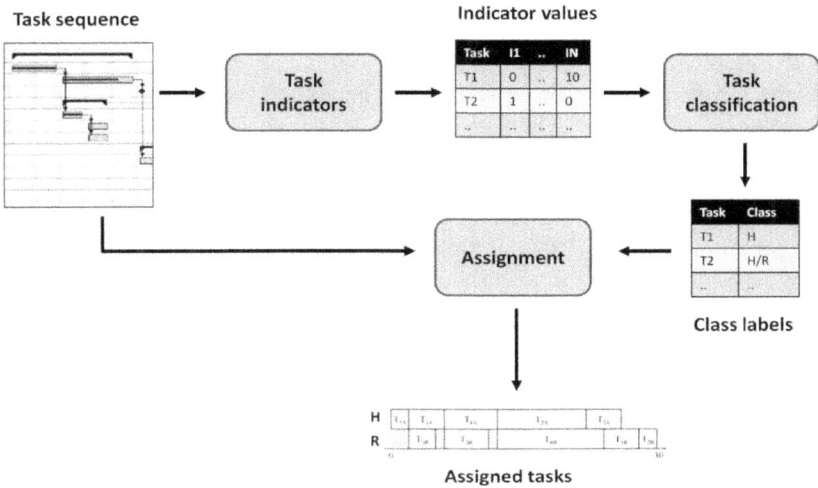

FIGURE 4.5 Procedure for task assignment to human and robot.

TABLE 4.2

Example of Indicator Values for Four Tasks

Task	W	Di	De	A
Tool retrieval	0	1	0	0
Inserting clamp	1	0	0	1
Welding	0	0	0	1
Fixing support	0	0	0	0

---

human, robot or collaborative work. There is also an important class consisting of the tasks that can be executed at will by a human or robot. The final step is task assignment to the first idle worker, respecting the constraints put on by the classifier. More details follow in the next section.

4.4.1 Task Indicators

The assignment of a task to a human or robot or the type of collaboration involved is a decision process. The decision process is supported by indicators that refer to the collaborative features of the task. That will be used as a decision factor in the selection of the type of collaboration. Having in mind the assets of humans and robots, some features need to be considered: the payload of the job (W), the length of displacement inside the work area (Di), the accuracy required to perform the task (A) and the necessity of executing dexterous moves or trajectories (De). Table 4.2 shows an example of application of the indicators to common tasks. The logical value of indicators in the example is 1 if the feature is present, that is, if accuracy is required, and 0 otherwise.

4.4.2 Task Classification and Assignment

ISO/TS15066 defines the safety requirements for collaborative industrial robot systems and for the environment in which they operate. Accordingly, there are different types of collaboration. The logic behind the definition of them is the presence of temporal and/or spatial separation between human and robot. The main cases of collaboration have been classified by ISO/TS 15066 as: "Safety-rated monitored stop (temporal and spatial separation), hand-guiding (temporal separation), speed and separation monitoring (spatial separation), power and force limiting (workspace sharing)". While sharing the workspace is a compelling collaborative goal, in most industries, speed and separation monitoring, or just hand guiding, is more widespread. Collaborative robots are not exploited to their best but just as an easy way to get rid of protective barriers. The reason is that full collaboration between human and robot would require a complete rethinking of productive/assembly tasks and perhaps a redesign of the task itself.

In the present example, the classes that have been defined are: to be executed only by the human (H), to be exclusively performed by the robot (R), to be executed at will by either the human and the robot (H/R) or to be the result of HRC (H+R). The supervised classifier was trained with the help of prior classified data, assigned by a team of experts. Continuing the example of Table 4.2, the aforementioned tasks are classified in Table 4.3. Supervised classification of logical data can be achieved by several machine learning algorithms. In this study, a C4.5 decision tree was used as the classifier (Quinlan, 2014), applying its open-source Java implementation that can be accessed in the Weka data mining tool (http://www.cs.waikato.ac.nz/ml/weka/).

The strategy of task assignment must be integrated by consideration of the task duration and precedence constraints other than just the classification. The most significant difference between the task assignment in HRC and conventional task assignment is that in HRC, it is not necessary to balance the amount of workload of every operator. As a matter of fact, a robot can and should work more and longer than its human teammate. In essence, the robot should take charge of risky, repetitive, tiresome tasks, unless human dexterity is required. The only tasks where a strategy is required is when the task can be executed at will by humans or robots (H/R). The number of such tasks depends on the kind of process considered: in assembly, a large number of parts can be handled either by robots or humans, while joining often requires ingenuity or at least dexterity.

TABLE 4.3
Example of Classified Data Used as Training Set

Task	W	Di	De	A	Class
Tool retrieval	0	1	0	0	H
Inserting clamp	1	0	0	1	H+R
Welding	0	0	0	1	R
Fixing support	0	0	0	0	H/R

4.5 PROGRAMMING COLLABORATIVE ROBOTS IN THE WORKCELL

Despite having witnessed strong growth in sales of robots to small or medium-sized companies over the last few years, there isn't a corresponding increase in the actual use of robots in production. The limiting factor continues to be the complexity and lengthiness of robot programming. Pan (2012) describes all the commonly used programming methods, either by online programming or offline programming, possibly with the help of VR. Both methods have seen consistent evolution in the last years. Online programming now can be performed without any knowledge of robots through sensor-assisted programming. Force sensors allow manual guidance of the arm of the robot: a pose is obtained by simply pushing and dragging the robot with a hand. Vision sensors allow a training-by-demonstration (TbD) approach: the programmer executes the task in front of the camera, and a governing system extracts a general policy for the robot trajectory. Manual guidance is already applied in some commercial robots: all Universal Robots, Kuka LBR and ABB YuMi. Vision-assisted TbD is still confined to research laboratories: movement policy generalization from demonstrations requires recourse to advanced machine learning techniques.

Billard (2008) complains that a large effort in TbD has been dedicated to studying "how to imitate" and "what to imitate," while less effort has been directed to studying "who to imitate" and "when to imitate." In other words, inadequate consideration has been given to the type and amount of experience the teacher should have and to the time instant chosen for the demonstration. In actual industrial processes, the importance of solving these last two issues becomes apparent. The robot should be trained to execute a welding task by observing a human welder with no experience in robot programming. The demonstration phase should waste a minimum amount of time and can be executed only when the parts to be welded are already fixed to the work table. Following Argall et al. (2009), industrial programming through TbD can be classified as derivation of a policy by a mapping function approach that applies regression on a dataset built by the external observation of the demonstration through remote sensors.

Consider the example of training the robot to execute welding presented in Antonelli (2013). The operator moves a light tool (named a pointer) along a given trajectory (Figure 4.6) to teach the robot the welding path. Some cameras observe the action and collect the poses of the pointer in an ordered list. From the list, the robot's trajectory is derived. Derivation of the path from the measured points is not trivial. Multiple overlapping segments are merged. Their points are properly reordered to obtain a single directed curve. The capability of the system to build a path from multiple overlapping segments is critical for industrial applications. There are many circumstances when indicating a path using just one continuous gesture is impossible or undesirable. Optical acquisition produces noisy data points, outliers that must be eliminated (Figure 4.7). The teacher's hand is not stable and is subject to wavering. The learning system should convert a fuzzy set of points in a welding trajectory to be performed with a robot's accuracy. The trajectory curve requires further smoothing and the elimination of excessive numbers of waypoints. The resulting smooth trajectory is shown in Figure 4.8.

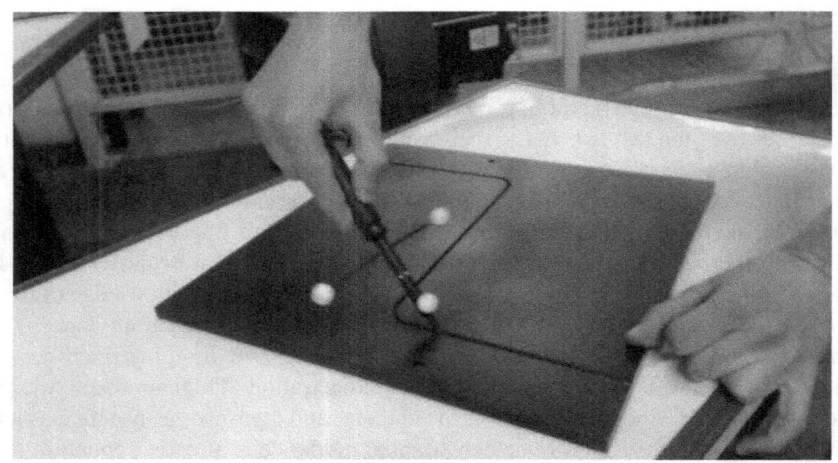

FIGURE 4.6   Robot training with training by demonstration.

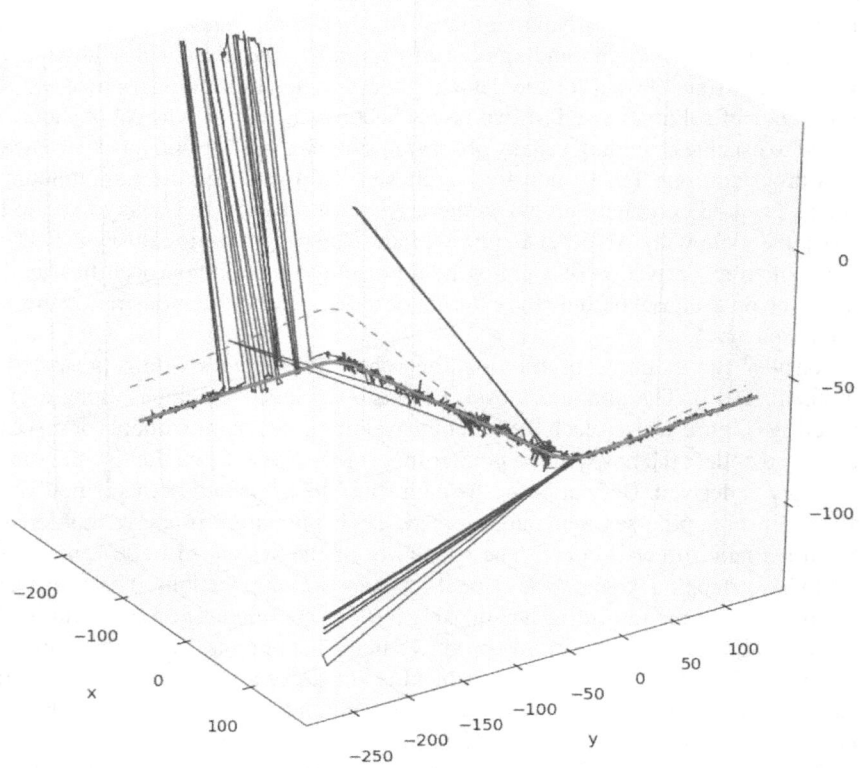

FIGURE 4.7   Noisy data acquired before outlier removal.

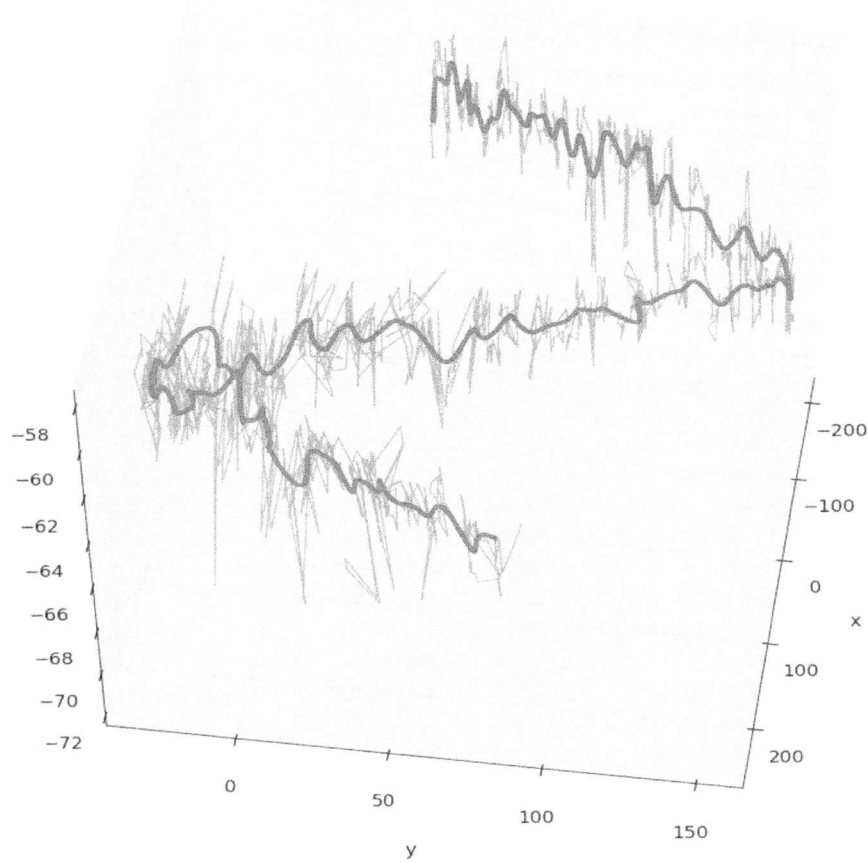

FIGURE 4.8   Resulting smoothed trajectory.

It is apparent that the trajectory is not yet optimal. The only way to obtain a feasible trajectory is to add in the system the knowledge of how the welding task should be executed. Welding is expected to produce a linear weld on flat surfaces. The curve of Figure 4.8 must be converted to a polyline by adopting an algorithm that reduces the number of intersected points, namely Douglas Peucker, obtaining the curve shown in Figure 4.7.

4.6   EXPLOITING COLLABORATION

Up to this section, no definition has been given for the term "collaboration". Collaboration has been applied to a wide range of human robot interactions inside the factory. Presently, in most industrial applications, it is used in place of sharing the same workspace. In Table 4.4, commonly accepted classifications of human–robot interactions are described.

TABLE 4.4

Types of Collaboration According to ISO/TS 15066

Time/Workspace	Spatial Separation	Spatial Coincidence
Temporal separation	Standard fully automated robotic workcell	Speed and separation monitoring
Temporal coincidence	Safety rated monitored stop	Power and force limiting

The maximum degree of collaboration involves the sharing of both time and workspace. Even so, interactions between human and robot are often limited to a few operations. It should be noted that the collaboration is not peer to peer, but it is more a coordination. Collaboration means working with the others for the success of an activity. Coordination is a kind of work optimization searching for the best division of activities to be performed by humans and machines. Following Malone (1994): "Coordination is managing dependencies between activities". Coordination implies the necessity of a coordinator and, therefore, a form of hierarchy. Instead, collaboration is the process of various individuals working together on a voluntary basis without the need for a manager or a work program. Collaboration implies the respect of a set of good behavior rules, such as the ones proposed by Talukdar (1999) in the case of collaborative software agents.

In the use of robot as an assistant for the human worker, that is, by carrying the greatest part of the workload, the task is apparently coordinated by the human, and the robot is employed as a useful assistant (Helms, 2002). It is not taken for granted that the future of HRC means new collaborative paradigms. With the evolution of the cognitive capabilities of robots, the usefulness of entrusting the robot with most of the coordination tasks will be evident. The robot knows the finished product perfectly thanks to access to computer-aided design (CAD) models. Robots access and combine the signals of a large number of sensors, always having accurate and real-time information about the current state of the work, and can detect possible nonconformities. For these reasons, robots could take the leading role in several assembly operations by indicating to the human operator the exact position and sequence of tasks to perform. Human workers would therefore be spared the task of keeping long and complex sequences of operations in memory. It would also lessen the need for expensive positioning devices and fixtures.

This way of collaborating with the robot means that the robot could be programmed offline: robot simulation software assists in generating trajectories and choosing the process parameters from the 3D-CAD model of the product. During production, the robot will have the burden of training the human operator to execute his/her part of the task, possibly with the help of AR. Regardless of who is in control, the success of collaboration relies on a prompt and reliable communication between the collaborative agents. Communication is reliable only if it is not one way. In human communication, even when only one is speaking, the other gives involuntary feedback of positive understanding by facial expression and eye movement. Replicating human-like feedback is complex and unnecessary; the robot could show what it has understood and what its next actions will be using virtual or AR simulations. The importance of communicating the intentions of the robot to the human is not neglected in modern research and even in some industrial applications (Mavrikios, 2013). Thus, the lack of empathy between man and machine makes it quite hard to correctly understand the exact intentions of the robot. This means that misunderstandings between human and robot will be frequent. It will be necessary to adopt mistake-proof procedures and devices and to quickly correct wrong actions.

4.7 CONCLUSIONS

This chapter discusses the topic of collaboration between humans and robots in industrial production processes. The main thesis presented in the chapter is that it is not enough to build a collaborative robot and meet the safety requirements. It is also necessary to redesign the workcell and also the methods of organizing the work in the establishment: the planning of operations and assignment of tasks. As for the implementation phase of collaborative work, the need for advanced and smarter robot programming methods is discussed. The chapter also highlights two fundamental aspects that should always be kept in mind when introducing an HRC-based production system: the need to steadily receive and give feedback messages to ensure correct communication and the importance of providing a system to correct misunderstandings during communication between human and robot.

REFERENCES

Almada-Lobo, F. (2016). The Industry 4.0 revolution and the future of manufacturing execution systems (MES). *Journal of Innovation Management*, 3(4), 16–21.

Antonelli, D., Astanin, S., Caporaletti, G., & Donati, F. (2014). FREE: Flexible and safe interactive human-robot environment for small batch exacting applications. In *Gearing up accelerating cross-fertilization between academic and industrial robotics research in Europe* (pp. 47–62). Springer.

Antonelli, D., Astanin, S., Galetto, M., & Mastrogiacomo, L. (2013). Training by demonstration for welding robots by optical trajectory tracking. *Procedia CIRP*, 12, 145–150.

Antonelli, D., & Bruno, G. (2017). *Ontology-based framework to design a collaborative human-robotic workcell*. 18th Working Conference on Virtual Enterprise, 18–20 September, Vicenza, Italy.

Argall, B. D., Chernova, S., Veloso, M., & Browning, B. (2009). A survey of robot learning from demonstration. *Robotics and Autonomous Systems*, 57(5), 469–483.

Bell, C. J., Shenoy, P., Chalodhorn, R., & Rao, R. P. (2008). Control of a humanoid robot by a noninvasive brain–computer interface in humans. *Journal of Neural Engineering*, 5(2), 214.

Billard, A., Calinon, S., Dillmann, R., & Schaal, S. (2008). Robot programming by demonstration. In *Springer handbook of robotics* (pp. 1371–1394). Springer.

Breazeal, C. et al. (1998). A motivational system for regulating human-robot interaction. *AAAI/IAAI*, 54–61.

Bruno, G. (2015). Semantic organization of product lifecycle information through a modular ontology. *International Journal of Circuits, Systems and Signal Processing*, 9, 16–26.

De Santis, A., Siciliano, B., De Luca, A., & Bicchi, A. (2008). An atlas of physical human–robot interaction. *Mechanism and Machine Theory*, 43(3), 253–270.

Fiorini, S. R. et al. (2015). Extensions to the core ontology for robotics and automation. *Robotics and Computer-Integrated Manufacturing*, 33, 3–11.

Goodrich, M. A., & Schultz, A. C. (2007). Human-robot interaction: A survey. *Found Trends Hum-Computer Interaction*, 1, 203–275.

Guarino, N. (1998). Formal ontology and information systems. In *Proceedings of the 1st international conference on formal ontologies in information systems* (pp. 3–15). IOS Press, Trento, Italy.

Hägele, M., Nilsson, K., Pires, J. N., & Bischoff, R. (2016). Industrial robotics. In *Springer handbook of robotics* (pp. 1385–1422). Springer International Publishing.

Haidegger, T., Barreto, M., Gonçalves, P., Habib, M. K., Ragavan, S. K. V., Li, H., Vaccarella, A., Perrone, R., & Prestes, E. (2013). Applied ontologies and standards for service robots. *Robotics and Autonomous Systems,* 61(11), 1215–1223.

Helms, E., Schraft, R. D., & Hagele, M. (2002). rob@ work: Robot assistant in industrial environments. In *Robot and human interactive communication, 2002. Proceedings* (pp. 399–404). IEEE.

Hinds, P. J., Roberts, T. L., & Jones, H. (2004). Whose job is it anyway? A study of human-robot interaction in a collaborative task. *Human-Computer Interaction*, 19(1), 151–181.

Kang, S. B., & Ikeuchi, K. (1995). Toward automatic robot instruction from perception-temporal segmentation of tasks from human hand motion. *IEEE Transactions on Robot Automation*, 11, 670–681.

Kazerooni, H. (1990). Human-robot interaction via the transfer of power and information signals. *IEEE Transactions on Systems, Man, and Cybernetics*, 20, 450–463.

Kim, S., Laschi, C., & Trimmer, B. (2013). Soft robotics: A bioinspired evolution in robotics. *Trends in Biotechnology,* 31(5), 287–294.

Klingspor, V., Demiris, J., & Kaiser, M. (1997). Human-robot communication and machine learning. *Applied Artificial Intelligence*, 11, 719–746.

Luh, J. Y. S., & Hu, S. (1998). Comparison of various models of robot and human in human-robot interaction. *IEEE International Conference on System Man and Cybernetics*, 2, 1139–1144. doi:10.1109/ICSMC.1998.727854.

Malone, T. W., & Crowston, K. (1994). The interdisciplinary study of coordination. *ACM Computing Surveys (CSUR)*, 26(1), 87–119.

Mavridis, N. (2015). A review of verbal and non-verbal human–robot interactive communication. *Robotics and Autonomous Systems,* 63, 22–35.

Mavrikios, D., Papakostas, N., Mourtzis, D., & Chryssolouris, G. (2013). On industrial learning and training for the factories of the future: A conceptual, cognitive and technology framework. *Journal of Intelligent Manufacturing*, 24(3), 473–485.

Nguyen-Tuong, D., & Peters, J. (2011). Model learning for robot control: A survey. *Cognitive Processing*, 12(4), 319–340.

Niles, I., & Pease, A. (2001). Toward a standard upper ontology. In *Proceedings of the 2nd international conference on formal ontology in information systems*. ACM.

Pan, Z., Polden, J., Larkin, N., Van Duin, S., & Norrish, J. (2012). Recent progress on programming methods for industrial robots. *Robotics and Computer-Integrated Manufacturing*, 28(2), 87–94.

Prestes, E. et al. (2013). Towards a core ontology for robotics and automation. *Robotics and Autonomous Systems*, 61, 1193–1204.

Quinlan, J. R. (2014). *C4.5: Programs for machine learning*. Elsevier.

Rosenstrauch, M. J., & Krüger, J. (2017, April). Safe human-robot-collaboration-introduction and experiment using ISO/TS 15066. In *Control, automation and robotics (ICCAR), 2017 3rd international conference on* (pp. 740–744). IEEE.

Talukdar, S. N. (1999). Collaboration rules for autonomous software agents. *Decision Support Systems*, 24(3), 269–278.

Wang, X. V., Kemény, Z., Váncza, J., & Wang, L. (2017). Human–robot collaborative assembly in cyber-physical production: Classification framework and implementation. *CIRP Annals-Manufacturing Technology, 66*(1), 5–8.

Womack, J. P., Jones, D. T., & Roos, D. (1991). *The machine that changed the world: The story of lean production.* Harper Collins.

Yang, J., Xu, Y., & Chen, C. S. (1997). Human action learning via hidden Markov model. *IEEE Transactions on Systems, Man, and Cybernetics—Part A: Systems and Humans, 27*, 34–44.

5 Machine Learning Applications for Industry 4.0

Vaibhav Shah, D.E.B. Costa, S.F. Moreira, J.F. Lima,
Maria Leonilde Rocha Varela and Goran D. Putnik

CONTENTS

DOI: 10.1201/9781003123866-5

5.1   INTRODUCTION

Historically, industrial revolutions have ridden on the engineering and technological advances of the day to increase production and supply capacity as well as creating new products that were not possible earlier. Each of the past industrial revolutions has fundamentally changed the way goods are produced, tested and distributed in terms of scale, efficiency and quality (of processes and products). While the first two industrial revolutions were trigged and powered by potent machines or hardware infrastructure, as seen in Figure 5.1, the third revolution was a digital revolution taking advantage of computing technologies. Further advancements in computing technologies, methodologies and required supporting hardware infrastructures have created favorable conditions for the 'fourth industrial revolution'. The 4th revolution, first discussed in the high-tech strategy by the German government in 2011, aims to utilize the current advances in cognitive and communication technologies and infrastructure, thus advancing from labor-intensive tasks to 'intelligent' tasks assigned to machines.

As in any revolution where sudden changes are noticeable and society naturally adapts, the 4th industrial revolution, termed Industry 4.0 (I4.0), has brought opportunities with the advancement of technologies that can facilitate, optimize and improve current industrial processes. Behind all industrialized products, there is a series of technologies involved and a chain of resources used from the beginning of the supply life cycle to the disposal of the product. According to Ohde and Mattar (2018),

> the extraction-manufacturing-use-disposal sequence remains the same for a large part of the industry, called the linear economy. In opposition to it, the circular economy

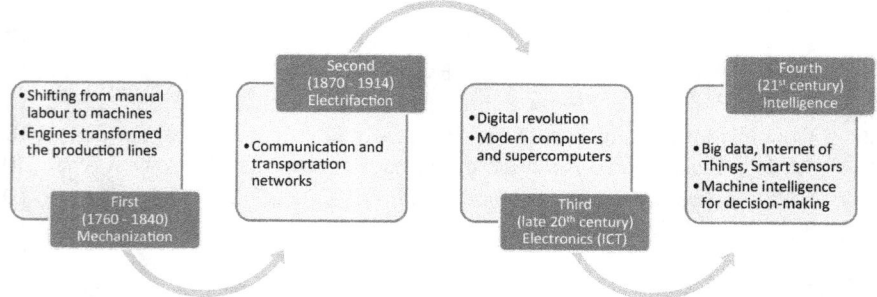

FIGURE 5.1   Industrial revolutions timeline.

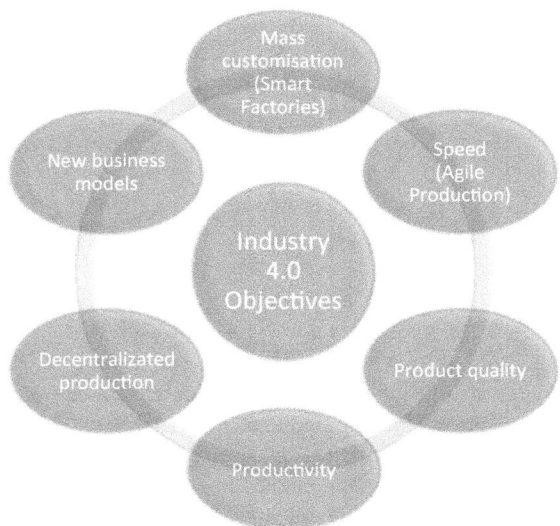

FIGURE 5.2   Industry 4.0 objectives based on the European Parliament briefing.
Source: Davies, 2015

was born, a process that has nature as its inspiration and deals with the natural and human resources that are prioritized in relation to capital and energy. It is a restorative economy, which is based on the reuse of the material that has already been extracted once from nature.

Current technological advances permit the reuse of materials and energy, instead of being discarded. This presents us with an opportunity to build a sustainable industry and planet. Besides sustainability, from the technological and process points of view, there are several major goals envisioned in the Industry 4.0 roadmap for productivity and growth (Davies, 2015), as shown in Figure 5.2.

The following technologies have been identified as the building blocks of I4.0 in the report (Rüßmann et al., 2015): autonomous robots, simulation, horizontal and vertical systems integration, the industrial Internet of Things (IIoT), cybersecurity, the cloud, additive manufacturing, augmented reality and big data analytics. Considering this, among major technological enablers are artificial intelligence–related technologies, such as machine learning (ML), robotics and industrial automation, supported by big data and cloud platforms. Moreover, one of the main characteristics of Industry 4.0 is a focus on the digitalization of physical assets and integration of digital systems, where digital interfaces play a major role in communication. In this sense, machine learning techniques enable the construction of the digital interface, assisting in automatic decision-making, especially in repetitive industrial processes, minimizing human interference that suffers fatigue over time. Various recent surveys have covered different aspects of integration of machine learning or related technologies in industrial processes. For example, some practical challenges as well as advantages of applying machine learning in manufacturing are discussed in Wuest et al. (2016), along with an overview of how to integrate ML algorithms and techniques in manufacturing processes. Two technological concepts discussed broadly in the context of I4.0 are IIoT and cyber-physical systems (CPSs), and a detailed recent survey presents different aspects of ML and big data analytics in this context (Xu et al., 2018).

There are strictly no rules about which part of the manufacturing process could or should be substituted or assisted by machine learning, but generally any process that requires decision-making or the expertise of trained personnel can be automated or improved by machine learning algorithms (Manupati et al., 2018; Putnik et al., 2021a, 2021b). It could be to assist a control system (Shah & Putnik, 2019), detect defects (Park et al., 2020) or tackle faults (Angelopoulos et al., 2020), another critical aspect for the success of Industry 4.0 as a technological revolution, besides discussing key aspects of machine learning applications in Industry 4.0. Considering this, the goal of this chapter is to present a brief idea of machine learning techniques and a practical guide to their implementation in industrial cases in the Industry 4.0 mission context. The objective is to give the reader a detailed idea of how to develop and deploy machine learning–based systems, integrate them in an industrial setup through two real-life cases and also, importantly, identify the advantages of such a system versus the same system without the assistance of machine intelligence. For this last part, an industrial case is considered where the difference in performance in quality control in a smartphone production plant with and without an automatic defect detection system is discussed.

The chapter is organized as following. Section 5.2 presents background on the main scientific topics, thus explaining the concepts of artificial intelligence (AI), machine learning and various learning techniques/algorithms. In Section 5.3, CRISP-DM methodology is explained as a guide for designing and deploying ML projects, with a complete project lifecycle for the ML part. Three real industrial cases of deployment of machine learning systems are presented in Section 5.4. Finally, the findings and conclusions are discussed in Section 5.5.

5.2 BACKGROUND

A detailed understanding of the concept of artificial intelligence requires knowledge of several disciplines, including computer science, neuroscience, psychology, mathematics, statistics and linguistics. The definitions of AI are mainly related to processes of thinking and reasoning of systems and machines, how their behaviors are addressed and how these dynamics are translated into processes. The applications of AI are now found in various areas such as robotics, computer vision, artificial neural networks and others, including machine learning. For a comprehensive understanding of the field, the authors recommend the classic book in the field *Artificial Intelligence—A Modern Approach* (Russell & Norvig, 1995). The term "machine learning" was initially proposed by Samuel (1959), suggesting building systems that could solve problems without being explicitly programmed for those. Another pioneer in the field was mathematician Alan Turing, who suggested shifting the problem from "can machines think?" to "can machines do what humans can do?", that is, tasks that require intelligence (1950). In other words, rather than putting effort into making 'thinking' machines, the focus should be on making machines that can do tasks that require a certain level of 'intelligence' and decision making.

Machine learning involves several disciplines that help in the integration of 'intelligence' into machines, data mining, statistics, pattern recognition, databases, data science and artificial intelligence. Machine learning is a branch of artificial intelligence aimed at making machines perform their jobs skillfully using intelligent software algorithms. Statistical learning methods are the lifeblood of intelligent software systems, which are used to create machine intelligence (Mohammed et al., 2016). In a simplified way, machine learning uses data that represents characteristics of the real world, which is processed by algorithms or learning models, with the ability to perform forecasting tasks, as shown in the following:

- Classification—Identifying which class/category a record belongs to.
- Estimation or regression—Like classification, but the record is identified by a numeric and not a categorical value.
- Prediction—Predicting the future value of an attribute.

These models undergo assessments of their ability to score correctly in relation to the forecast. Mohammed et al. (2016) highlights that the model can be considered an approximation of the process that we want machines to imitate. Sometimes we can understand the model and sometimes not, characterizing the latter as a black-box system, where its operation cannot be explained intuitively.

5.3   MACHINE LEARNING METHODS

Figure 5.3 presents four principal machine learning approaches related to input data, with a brief description.

Supervised learning is that in which there is a training data set labeled with correct answers for each input sample for learning. *Unsupervised learning* is that in which there are no labels for learning. *Semi-supervised learning* is that in which only a few data points are labeled. Reinforcement learning is the generation of machine learning models to make a sequence of decisions in a potentially complex environment (Data Science Academy, 2019). In Figure 5.4, these approaches are divided into sub-branches (Ribeiro, 2019).

5.3.1   SUPERVISED LEARNING

In supervised learning, the objective is stated with labeled/classified training data. There is an output/target value, an assigned class, for each input/training sample. Thus, the training data consists of the input vector, say, X, and a corresponding output vector, say, Y (or labels), thus establishing the relationship function $f(X) = Y$. Together, they form a training example that can describe a relationship that 'makes sense'. For example: Parents teach their children about the names (labels) of objects by pointing at them and pronouncing their names in a supervised manner.

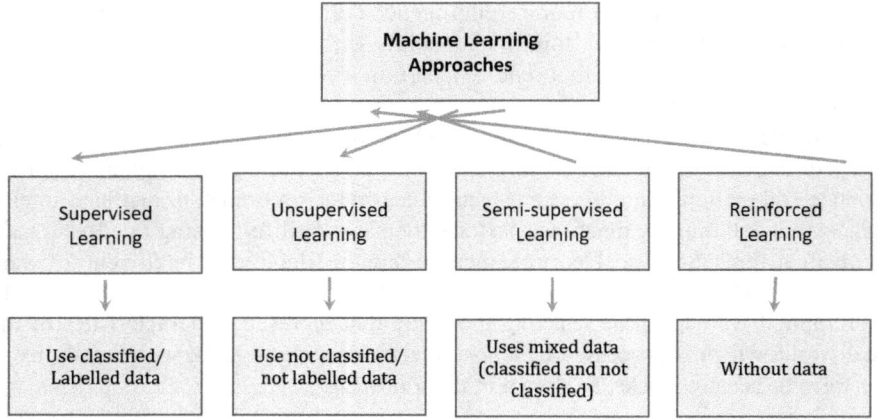

FIGURE 5.3   Machine learning approaches and their required data.

Source: Mohammed et al., 2016

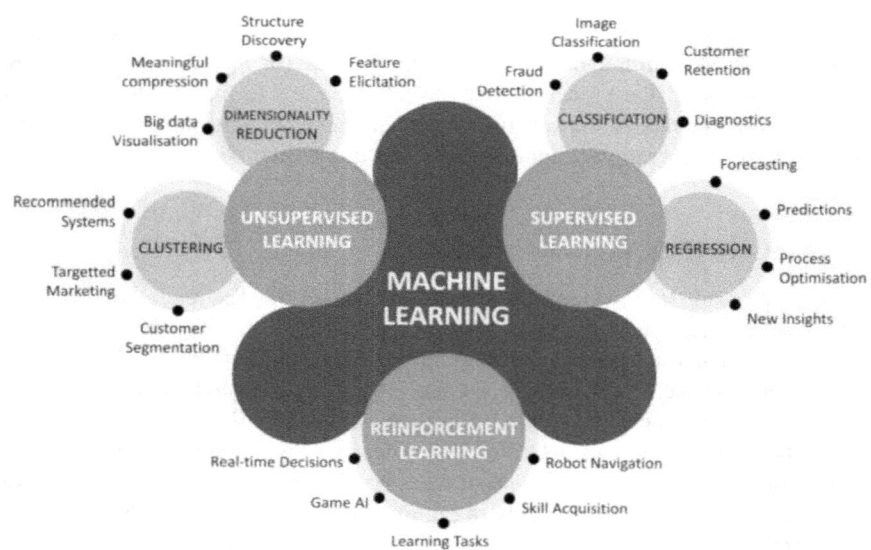

FIGURE 5.4   Machine learning in its three paths.

Source: Adapted from Ribeiro, 2019

5.3.2   UNSUPERVISED LEARNING

In unsupervised learning, there is no supervisor indicating/labeling a vector with a defined objective, and therefore there is no 'labeled' data for training. Sometimes there may not be total clarity about what is to be learned. This is a kind of situation data scientists and machine learning engineers often find themselves in, when a large amount of data is available but without preassigned labels. Still, it is possible to learn by searching for significant or hidden structures or patterns in the data.

5.3.3 REINFORCEMENT LEARNING

Reinforcement learning is the training of machine learning models through a sequence of decisions. The 'agent' (being trained) learns to achieve a goal in an uncertain and potentially complex environment while receiving positive or negative feedback for each situation. The computer uses the trial-and-error method to find a solution to the given problem, receiving 'rewards' for each decision taken by the system. The goal is to maximize the total rewards. We can see the flow of a feedback loop in Figure 5.5.

Following are brief descriptions of some of the main algorithms used in machine learning applied for building a model that describes a function of the relationship between predictor variables (X or attributes) as inputs and predictive variables (Y or classification or results). Although a large number of algorithms are applied in practice currently, as seen in Figure 5.6 adapted from Gonçalves (2019), four of them are described briefly in the following.

5.3.4 DECISION TREE

Côrtes et al. (2002) define a decision tree as a flowchart like a tree structure, where each node represents a test on a given attribute, while each branch (sub-tree) represents the test result and each leaf represents the distribution of records. They are the most-used models in inductive inference, where the model is trained according to a set of previously classified examples, which characterizes this as supervised learning, allowing future classification or prediction based on the modeled tree.

5.3.5 SUPPORT VECTOR MACHINES

A support vector machine (SVM) is another supervised algorithm used for pattern recognition and classification tasks. According to Côrtes et al. (2002), the separation

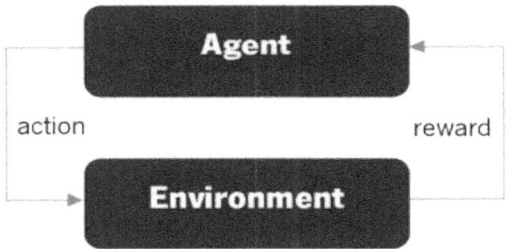

FIGURE 5.5   Reinforcement learning process.

Source: Gonçalves, 2019

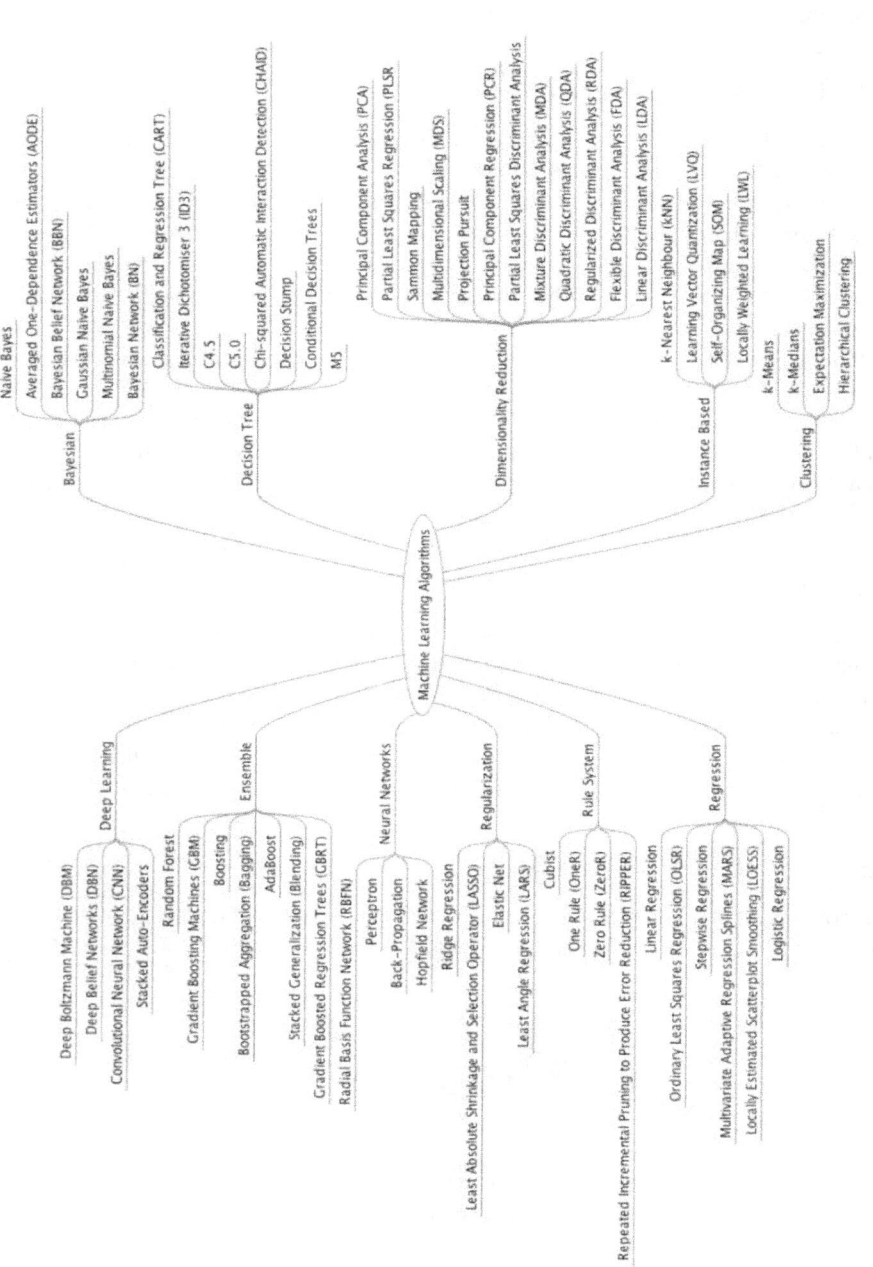

FIGURE 5.6 Machine learning algorithm representation.

Source: Gonçalves, 2019

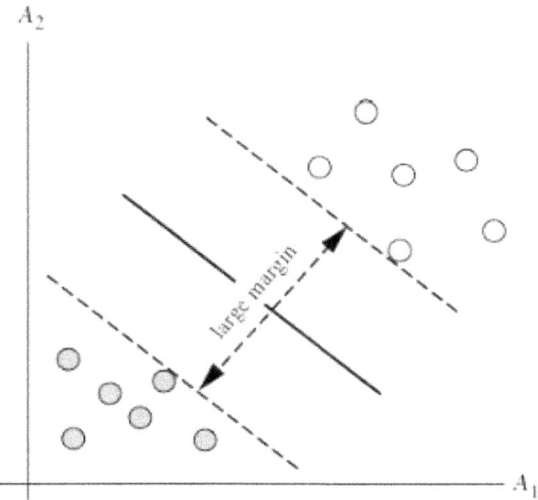

FIGURE 5.7   Training data where there is an infinite number of hyperplanes that can separate the classes.

Source: Adapted from Côrtes et al., 2002

of classes is done by a hyperplane that uses support vectors (training set), making the separation of groups visible. Figure 5.7 shows an example of training data classified as "below the goal" (gray balls) and "above the goal" (white balls), where the algorithm seeks to identify groups by the presence of the hyperplanes that separate the classes.

It is one of the machine learning algorithms used for classification, where the response variable is categorical. In a summary made by Lin (2018), SVM models work by predicting the probability of Y belonging to a specific category, first adjusting the data to a linear regression model, which is then passed on to the logistic function. The logistic function will always produce an S-shaped curve, so, regardless of X, a sensible response can always be obtained (between 0 and 1). If the probability is above a certain predetermined limit [for example, P (Yes) > 0.5], then the model will predict Yes.

5.3.6   NAIVE BAYES

Naive Bayes is also a supervised learning method, using algorithms based on the application of Bayes' theorem. Camilo & Silva (2009) describe Bayes' theorem as: "it is possible to find the probability that a certain event will occur, given the probability of another event that has already occurred: Probability (B given A) = Probability (A and B)/Probability (A)". The naive Bayes algorithm assumes that there is no dependency relationship between the attributes.

5.4   CRISP-DM: METHODOLOGY FOR ML APPLICATIONS IN INDUSTRIAL PROCESSES

The most common methodology for data mining, or implementing machine learning or data science projects in production, is called the CRISP-DM technique, which is presented in this section. It is a cyclical process model with six phases that describe the data science project life cycle, like a set of guard rails to help plan, organize and implement a data science (or machine learning) project in order to make the process of development of machine learning in industrial processes easier to deploy, evaluate and readjust. It was published in 1999 to standardize data mining processes across industries (Shearer, 2000). According to Jeff Saltz's survey in August and September 2020, conducted among several organizations, 49% of them are using CRISP-DM compared to other method, like Scrum (18%), Kanban (12%), My Own (12%), TDSP (4%) and others (6%).

The steps are shown in Figure 5.8. As seen in the figure, the steps are executed in a cyclical manner; that is, once the evaluation is performed, the experts may go back to any of the previous steps to readjust one or more parameters or even re-evaluate their business understanding. Additionally, for detailed understanding, the steps are explained in Table 5.1.

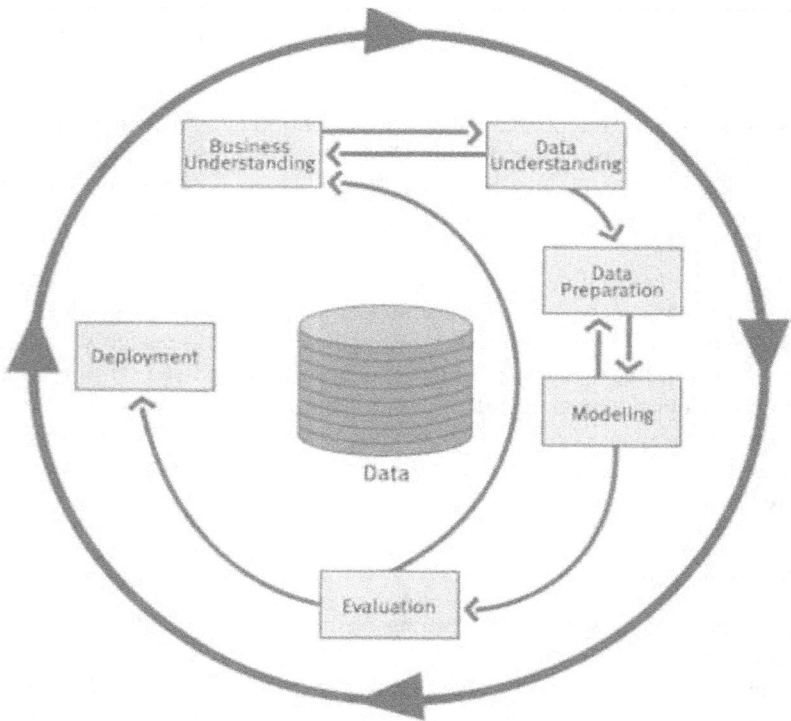

FIGURE 5.8   CRISP-DM summary flowchart.

Source: Adapted from Shearer, 2000

TABLE 5.1

Steps of CRISP-DM. Adapted From Data Science Project Management (1999)

Steps	Description	How to Apply the Steps
1	**Business understanding** What does the business need?	1—Determining the business objectives: "thoroughly understand, from a business perspective, what the customer really wants to accomplish" (CRISP-DM Guide) and then define business success criteria. 2—Situation assessment: Determining resource availability, project requirements, risk assessment and contingencies and conducting a cost-benefit analysis. 3—Defining success from a technical data mining perspective. 4—Project plan creation: Selecting technologies and defining detailed plans for each phase.
2	**Data understanding** What data do we have/need? Is it clean?	1—Primary data collection: data acquisition, as well as feeding into the tool. 2—Data description: Examining data and registering data properties such as format, number of samples (data volume) and so on. 3—Data exploration: visualizing data and identifying relationships. 4—Data quality check: Analyzing the noise among data and noting quality issues.
3	**Data preparation** How do we organize the data for modeling?	1—Selection: Selecting the datasets to be used and registering the rationale behind the selection/exclusion of datasets. 2—Data cleaning: Might be a laborious task but is nevertheless critical for the quality of model training to avoid errors at the training level. 3—Data construction: Creating new attributes, such as calculating body mass index (BMI) from weight and height values. 4—Data integration: Generating new datasets through integration. 5—Data re-formatting: Converting data from one format to another for ease of operations.
4	**Modeling** What modeling techniques should we apply?	1—Algorithm selection. 2—Test designing/planning: preparing datasets for training and testing, for hypothesis validation and so on. 3—Model generation: The actual task of training the model. If implementing through a high-level programming language such as Python, this might be as simple as "reg = LinearRegression().fit(X, y)". 4—Model assessment: Typically, several models are trained, and the data scientist or ML expert interprets the results with the help of domain knowledge and previously agreed-upon criteria for success and hypothesis.
5	**Model Evaluation** Which model achieved the business objectives best?	1—Results evaluation: Have the model(s) met the success criteria? Selecting the most promising model(s) for the business. 2—Reviewing: Critically analyzing the tasks completed to check whether something was skipped or if everything was executed correctly. As a result, correcting anything if required. 3—Deciding the next steps: Building upon the prior tasks, deciding about the next steps—either deployment, re-iterate further or reinitiate.
6	**Deployment** How do stakeholders access the results?	1—Deployment: Preparing a plan for model deployment into production. 2—Monitoring and maintaining the plan: In order to avoid operational phase issues of a deployed model. 3—Final project report: A final detailed project summary is prepared by the people involved in the previous phases. This should include model evaluation results. 4—Project review: A review and analysis of the positives and negatives of all the steps and what and how to improve in the next or future cycles.

For sustainable manufacturing, machine learning can be applied to achieve the targets in the form of key performance indicators (KPIs).

5.4.1 STEP 1: BUSINESS UNDERSTANDING

In this step, the business is contextualized to understand how machine learning techniques and applications can be helpful, considering the big picture and looking in a non-technical way. Shearer (2000) emphasizes that this is an important phase of the process because through this, it is possible to develop a concise plan for the acquisition of the data aligned with the business objectives. The problem, objectives and all the necessary information from the end-user's perspective are gathered and evaluated.

5.4.2 STEP 2: DATA UNDERSTANDING

After the business understanding, that is, understanding the problem and its strategic and tactical objectives, the next phase is observation of the relevant data. For Shearer (2000), in this phase, familiarization with the available data and the first experiments, as well as evaluation of the quality of the data and the first conclusions and initial groupings, are important in order to avoid problems and advance more assertively to the next step.

5.4.3 STEP 3: DATA PREPARATION

This phase encompasses all tasks to prepare the final dataset or the input data that goes into the modeling tool/algorithm from the initial raw/unprocessed data. The main steps in data preparation are: data selection, data cleaning (noise reduction) and feature selection or extraction, and, depending on the case, there may be a need to either reduce/eliminate features (dimensionality reduction) or create new features. There are multiple approaches for this last process, including techniques such as principal component analysis (PCA). A good summary of data preparation steps and techniques is provided in Roh et al. (2019).

5.4.4 STEP 4: MODELS/ALGORITHMS OF MACHINE LEARNING

In this phase, one or more modeling techniques/algorithms are selected and applied, and their parameters are configured to values considered optimum. Usually, multiple techniques can be applied for the same data mining problem. Some of them have particular requirements regarding the data type or format. Hence, it may be necessary to step back to data preparation. Quoting Shearer (2000), "Modelling steps include the selection of the modelling technique, the generation of test design, the creation of models, and the assessment of models". After deciding the modeling technique, the test design must be generated. This is ensured by the data analyst by testing for the model's quality and validity, eventually carrying out empirical testing in order to determine the model's strength. Typically, the datasets are divided into training (75%) and testing (25%) datasets, or other percentages according to the project requirements.

Once the algorithm/approach is decided upon, the next step is to choose the machine learning tool or tools to implement the system. There are two possible ways: 1) implement the algorithm using a programming language or software development framework like C++, Python, R, Java and so on or 2) build and test the machine learning model using a ready-to-use ML tool. Ribeiro et al. (2018) have studied some machine learning technologies (tools) in detail in order to understand which are the most used tools in the scientific community. The main technologies analyzed were: SparkMLlib, TensorFlow, Scikit-Learn, Weka, Dask, Caffe, Microsoft Cognitive Toolkit, MXNet, Accord.NET, Theano, PaddlePaddle, DeepLearning4j, Apache Singa, Apache Mahout, KNIME, VowpalWabbit, R, Keras, XGBoostPytorch, RapidMiner, OpenCV and Chainer. Another data mining toolbox popular among the research community is the Orange tool, developed by a team of researchers from the University of Ljubljana (Demšar et al., 2013). After selecting the algorithm and tool, the algorithm is applied on the training dataset and a 'model' or 'trained model' is generated. After generating the model(s) through the selected machine learning algorithm(s), it is necessary to evaluate the performance of the trained model(s), as is shown in the next step.

5.4.5   Step 5: Evaluation

To evaluate the performance of the model, some indicators are considered related due to their ability to estimate, predict or classify new instances given a set of training and test data. The confusion matrix shows the predicted and real classifications by a machine learning model, as presented in Kohavi and Provost (1998). Visually it provides the hits and misses of the model, according to its impact on the processes in which they are inserted—manufacturing decision processes or disease diagnosis, for example. Table 5.2 shows the confusion matrix of a binary classification algorithm (yes/no, sick/healthy, defective/non-defective, positive/negative, belongs/does not belong), where the prediction of the model is compared with the actual value. This matrix has four possibilities for classifying the model's output.

- True positive (TP)—Model predicted "Yes" and real result "Yes".
- True negative (TN)—Model predicted "No" and real result "No"
- False positive (FP)—Model predicted "Yes" and real result "No"
- False negative (FN)—Model predicted "No" and real result "Yes".

TABLE 5.2

Binary Classification Confusion Matrix. Adapted from Kohavi and Provost (1998)

		Predicted	
		Yes	No
Actual	Yes	True positive	False negative
	No	False positive	True negative

The correct predictions of the model are on the diagonal (TP and TN), and the errors are outside the diagonal (FP and FN). With the confusion matrix, it is possible to obtain the typical metrics used to evaluate ML models. The meanings and formulas of the metrics are briefly presented in the following, based on Matos et al. (2009):

- **Accuracy**—Measures the ability of the model to correct its predictions; that is, it measures the fraction of correct predictions (TP and TN). This metric is not appropriate when the classification of instances for training is disproportionate and a lot of information is irrelevant, in other words, class imbalance.

$$Accuracy = \frac{\left(TP + TN\right)}{\left(TP + TN + FP + FN\right)} \tag{1}$$

- **Precision**—Shows the model's ability to correctly classify/predict as positive; that is, it measures the fraction of true positives in relation to all cases that are predicted to be positive.

$$Precision = \frac{\left(TP\right)}{\left(TP + FP\right)} \tag{2}$$

- **Recall**—A rate that evaluates the model's ability to classify/predict as positive what is really positive; that is, it measures how many actual positives are predicted as positive.

$$Recall = \frac{\left(TP\right)}{\left(TP + FN\right)} \tag{3}$$

- **% False Positive**—The percentage of errors when classifying negative instances as positives.

$$\%FP = \frac{\left(FP\right)}{\left(TP + TN + FP + FN\right)} \times 100 \tag{4}$$

- **% False Negative**—The percentage of errors when classifying positive instances as negatives.

$$\%FN = \frac{\left(FN\right)}{\left(VP + VN + FP + FN\right)} \times 100 \tag{5}$$

The various machine learning models are submitted to accuracy, precision, recall, % FP and % FN evaluations, where the best performances lead to the choice of the best model. This step requires an understanding of the errors and successes of the model to improve performance. If it is necessary, come back to step 2 and run the cycle again.

5.4.6 Step 6: Deployment

In industry, this is the testing and integration phase of intelligent software in the process for which it was created, considering its industrial application, such as:

- Equipment for data capture (quantities and position in the process)
- Automation
- Communication protocol with manufacturing execution system/integration with IT
- Operation procedure of the new process
- Batch pilot and validation
- Production with new test/process
- Communication and training of those involved
- Revisions needed

At this stage, it is important that the innovations and benefits of the initially defined project be demonstrated.

5.5 PRACTICAL APPLICATIONS IN MANUFACTURING PROJECTS

In this section, three ML project application examples for problem solving in industry are presented as real-life use cases. These projects showcase different aspects of implementing and integrating machine learning techniques in industrial setups in order to help readers in their own machine learning project implementations.

5.5.1 Example 1: Machine Learning to Inspect Refrigerant Gas Leakage in Air Conditioners

In this example, the methods of data collection, preparation and training are described by adopting a supervised machine learning technique with algorithms that allow the evaluation of their performance in classifying leakage and non-leakage regions in a product using Orange.

5.5.2 Business Understanding

In an air conditioning manufacturing process, one of the most critical points for maintaining product quality is guaranteeing that the refrigeration system is sealed correctly, with no holes that allow refrigerant gas to leak. Several joints between copper tubes are part of the product structure, necessary to close the system through which the gas circulates, as shown in Figure 5.9. These joints are made through an alloy deposited in the joints through the brazing process. These are the most vulnerable positions to a gas leak. The absence of refrigerant gas inside the refrigeration system causes the air conditioning to fail to cool or heat the environment.

In the current process workflow, there are two gas leakage tests using high-pressure methods in the refrigeration system, varying only in the type of gas. The effectiveness of this test depends on the performance among measuring equipment;

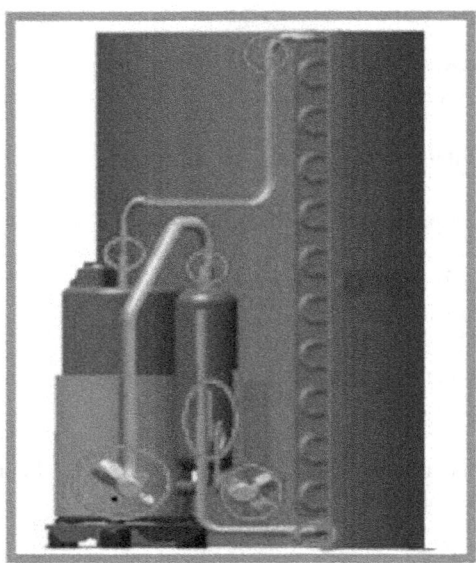

FIGURE 5.9 Condensation system and brazing joints points in a heat exchanger.

product under test; and operator, who has an important role in guaranteeing the test method (approximation of the welding points and the speed of the probe and the equipment). Even with all the current tests in the process, there are still products reaching the consumer's home without refrigerant gas or with insufficient gas. Proposed solution: This study looks for alternatives that improve detection, reducing human interference in the judgment, given the speed of industrial production, with low area occupation. A thermal camera was used to capture a thermographic image to study temperature patterns in leaky and leak-free regions of the heat exchanger. As this is an undeveloped technology, a concept test was adopted to analyze the project's feasibility.

5.5.3 Data Understanding and Preparation

To learn about the behavior of the surface temperature of the exchanger, initial tests were carried out on air-conditioned products with two leak sizes on the order of 3g/year, using an Optris 640 model camera, with R410 gas inside the product. Eight tests were carried out, with temperature mappings in the leaking region and non-leaking regions. For each test, a video was recorded with the thermal camera, recording the temperature of the object present in the recording for a period of 20 seconds. Software that comes with the camera makes it possible to view the video and export the temperature data selected in the image to an Excel spreadsheet. Eight points (P1 to P8) were selected for temperature collection.

Table 5.3 shows a sample of the training table, with the attributes in the columns and the instances in the rows.

TABLE 5.3
Training Data Set Sample

	Product	(A) Size of Leakage	(B) Blanket	(C) Camera Distance	Classifi-cation	Meas. Position	Temp. (t 0)	Temp. (t 0,2)	Temp. (t 0,4)
Instances	B	0.12mm²	sem	10cm	Leakage	P1	27.4	27.4	27.3
	B	0.12mm²	sem	10cm	Normal	P2	27.9	28	27.8
	B	0.12mm²	sem	10cm	Normal	P3	28.2	28.2	28.2
	B	0.12mm²	sem	10cm	Normal	P4	28.1	28.1	27.9
	B	0.12mm²	sem	10cm	Normal	P5	27.9	27.9	27.8
	B	0.12mm²	sem	10cm	Normal	P6	30	30	30.1
	B	0.12mm²	sem	10cm	Normal	P7	27.4	27.3	27.2
	B	0.12mm²	sem	10cm	Normal	P8	28.9	28.9	28.9
	A	0.22mm²	com	10cm	Leakage	P1	28.8	28.9	28.7
	A	0.22mm²	com	10cm	Normal	P2	28.5	28.5	28.5

TABLE 5.4
Total Data for Training and Testing the Algorithms

	Training Data		Testing Data		
	Normal	Leakage	Normal	Leakage	Total
Instances	28	4	28	4	64

The total set of data for training and testing the algorithms is summarized in Table 5.4. It should be noted that the data has no balance between the instances classified as "Normal" and "Leakage". This was due to the limited number of samples for the study.

5.5.4 MODELING AND ALGORITHM EVALUATION

Figure 5.10 shows the machine learning study using the Orange software, with tools ready to be processed. The study was built in three blocks: Block 1, data loading; Block 2, learning algorithms; and Block 3, performance evaluation of the algorithms based on the input data (test and training).

Four supervised learning algorithms (decision tree, logistic regression, SVM and naïve Bayes) were applied in this case. To perform the evaluation of the algorithms and respective models, the tool "Test and Score" was used, with input data (training and test tables) and the algorithms for training performance evaluation. The four algorithms were evaluated by the confusion matrix and performance indicators that measure the ability to predict by classifying whether it is a region with "Leakage" (TP) or "Normal" (TN). Since the objective is to detect whether there is any possible leakage, a sample with leakage is considered positive if leakage is found and negative if there is no leakage found.

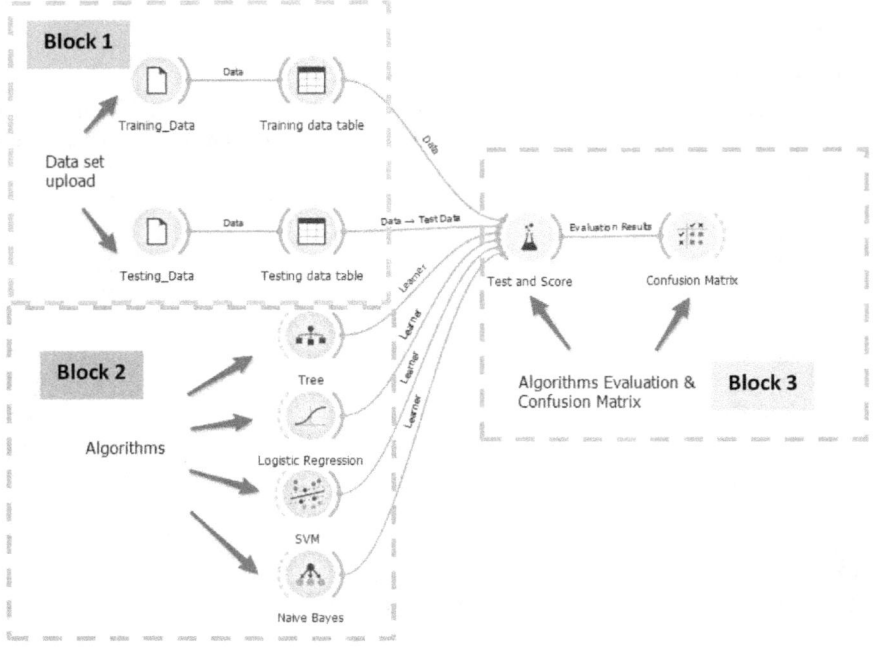

FIGURE 5.10   Machine learning project for leakage classification.

TABLE 5.5
Confusion Matrix of Results from Models Trained by Different Algorithms

Algorithms and Their Predicted Outcomes vs. Actual Labels		"Normal"		"Leakage"		Total
		True Negative	False Positive	True Positive	False Negative	
Predicted	Tree	27	1	0	4	32
	SVM	28	0	1	3	32
	Logistic regression	28	0	2	2	32
	Naive Bayes	12	16	2	2	32
Actual		**28**		**4**		32

5.5.5   Validation—Algorithm Performance

The predictions by the models generated by four algorithms are presented in a combined confusion matrix in Table 5.5. Predictions of the classification by each algorithms, as compared to the test data, are categorized into four types of forecasts:

- True positive: Predicted with "Leakage" and actually with "Leakage";
- True negative: Predicted without "Leakage" and actually without "Leakage";

TABLE 5.6

Evaluation of the Algorithms/Models

Algorithms	Higher Is Better			Lower Is Better	
	Accuracy	Recall	Precision	% FP	% FN
Decision Tree	0.84375	0	0	0.03125	0.125
SVM	0.90625	0.25	1	0	0.09375
Regression	0.9375	0.5	1	0	0.0625
Naive Bayes	0.4375	0.5	0.111	0.5	0.0625
Interpretation	Total % samples classified correctly	% actual defects detected out of all actual defects	% actual defects out of all samples classified defected	% False positives	% False negatives

- False positive: Predicted with "Leakage" and actually without "Leakage";
- False negative: Predicted without "Leakage" and actually with "Leakage".

The test data consists of 28 instances of areas with no leak or "Normal" (TN) and 4 instances of areas with "Leakage" (TP), that is, 'areas of interest'. With this complementary analysis using the confusion matrix, it is possible to observe that the decision tree algorithm classified the four "Leakage"s as false negative. The naive Bayes classified 16 false positives, where both had a low performance in classification. Logistic regression correctly classified all true negatives and hit 50% of true positives.

Subsequently, based on the values in the confusion matrix, evaluation metrics are calculated to measure performance by each algorithm model. Table 5.6 shows a comparison of the performance results of the trained models considering the evaluation indicators.

Following are major observations from this evaluation process:

- Out of the four algorithms evaluated, the decision tree had the worst result in precision, recall and F1 metrics. This was followed by naive Bayes, with a high FP index, that is, the rate of normal products, without leakage, being identified as defective, or leaky. In this case, 50% of all samples were falsely identified as defective.
- Logistic regression showed better results among the four: 100% precision in TP classification and zero FP classification. However, there is scope for improvement in the recall indicator, with 50% of correct answers for actual defects (TP), affected by the FN classification.
- It is important to note that more samples are needed for a more precise calculation and to increase confidence in forecasts, meaning more air conditioning process data would be needed, especially those with "Leakage" cases, so that the accuracy assessment uses balanced TP and TN data. Currently, there is a huge imbalance in the two types.

5.6 CONCLUSIONS AND FURTHER RESEARCH

During the ongoing research, it was possible to observe the evolution of knowledge about the use of an infrared thermographic camera to identify gas leaks in an air conditioning system within 20 seconds and using the surface temperature data of the product captured by the camera for supervised machine learning, classifying a region either as "with leak" or "normal". The open-source software Orange was used for machine learning and data analysis, which allowed the evaluation of algorithms. In the experiments on the available data, the logistic regression algorithm obtained better classification performance, with opportunities for improvement in future work with more data collection than the available data at the time of writing.

Nevertheless, this work demonstrated that it is possible to automatically detect gas leakage defects in the real case/environment/industrial setup, which is the main contribution of this work. It provides validation of a new concept of leakage tests of gas in air conditioning/refrigeration systems, providing security for the company to move forward with the development phase of software, hardware and integration of the camera into the production process, with the purpose of transferring the judgment/detection of whether it is a "Leaky" product or "Normal" to an automatic intelligent system.

5.6.1 EXAMPLE 2: AI-ASSISTED DIGITAL IMAGE PROCESSING FOR METROLOGY SYSTEMS

In this section, an image processing system as a solution for a statistical process control improvement is presented. This process is important to ensure finished good quality by measuring samples of data from production line for specific characteristics to reduce cycle time of operators going to the production shop floor to collect paper sheets, then return to the office and manually input information into a repository database. Additional improvements are related to increasing data reliability, avoiding missing registration or registering wrong information by operators. Therefore, an image measurement system with AI resources can be implemented to achieve these targets.

5.6.2 BUSINESS AND DATA UNDERSTANDING

For heating, venting and air conditioning (HVAC) system manufacturers, heat exchange is a critical part. So, it is important that quality is ensured to maintain customer satisfaction. This is done by measuring some key characteristics (KCs) through the production process. In the case of the components that must ensure good heat transfer from inside a room to outside a room, the product must ensure thermal performance without leakage. Key characteristics to ensure this are quantity of frames per inch (FPI) to ensure the air flows through the copper tubes in right the amount and the diameter and height of coil nozzles to keep the correct assembly of copper tubes before the brazing process and avoid leakage issues. These KCs can be measured manually by an operator, but it will take up valuable resources in factory. A good alternative to make this process agile is to develop a metrology system that acquires product images and shows the results of KC measurements, as demonstrated in the next sections.

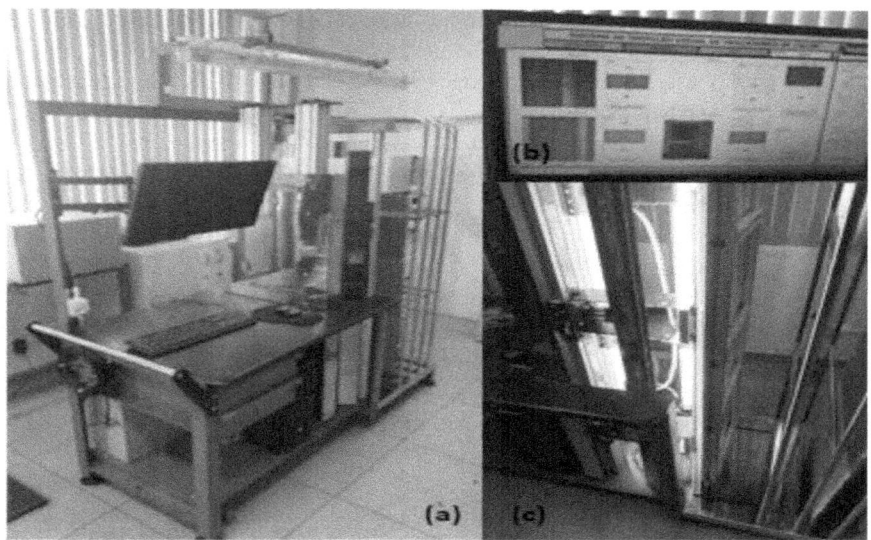

FIGURE 5.11   Detail of (a) equipment for measuring FPI and heat exchanger nozzles; (b) visual inspection system integrated into the system; (c) product placement location for image acquisition.

5.6.3   System Design (Data Preparation)

The metrology system was designed to do image acquisition, making a comparison with a known standard so that the main information is obtained. In the case of FPI measurement, height and collar diameter, the software must have pattern detection requirements (in this case, edge detection) to accurately obtain the components' dimensions. For the patterns to be captured, the product must have a receptacle for its temporary packaging in a fixed and immobile position, and the chambers and lighting must be correctly positioned, according to the details in Figure 5.11.

In the figure, it is also possible to see the visual inspection and data acquisition system. In addition, the system must have the ability to store the measured results in a database for further analysis. The equipment described previously was developed by a supplier according to the manufacturer's requirements and specifications. Validation of the measurement system is also required, as will be seen in the next sub-section.

5.6.4   System Validation

Metrology systems must have their precision evaluated. Measurement errors must be identified and evaluated. These measurement errors are caused by the process itself. Factors like methods, equipment and operator skills can influence measurements, so it is very important to understand which factor(s) most influence measurement errors. Therefore, it is necessary to carry out an analysis of the measurement system (MSA) with the collection of measurement and analysis data through control charts. For the measurement of FPI, measurement samples were collected with the new system, and the new system was compared with the instruments already used, considering factors of process variation. Sampling trees were made to plan FPI measurements and mouthpieces, as shown in Figure 5.12.

For the MSA analysis, three components were selected, which were measured by two operators, each operator making three measurements on each of the instruments. These three measures were to be used for the formation of subgroups, and in each of these, the average and the variation between the elements were calculated with the objective of building the X-bar R charts and calculating the LSC and LIC. The measurement values for the height and diameter of the mouthpieces were collected and are shown in the following figures, with the comparisons between the new system and the measurement made with a caliper. It was also found that the pattern of the X-bar charts was repeated in the two measurement systems (showing that the new system has reproducibility) and that most of the measurements were outside the LSC and LIC (indicating that the variation of the measurement system is less than the variation of the process itself).

On the other hand, the R chart shows us that there were variations outside the control limit, indicating that there were variations between measures above the expected variation of the process, which means there is no repeatability in the system. Therefore, the system is not approved, and the causes of variation in the measurement system will need to be reviewed. In the charts in Figure 5.13, it is possible to identify that the image inspection system showed a greater variation in the control limits than the measurement with the current instrument.

Together with the equipment supplier and the departments directly involved in the application and based on the results collected from the MSA 1, several improvements were proposed in the FPI bench. These include activation of the lighting system, installation of supports and guides so that the part is as secure as possible and as straight and fixed as possible, changing the bases to avoid tilting the product, making product supports to avoid changing the position of the chambers and making physical protection for the chambers. The software has also been analyzed and improved. After the changes were applied to the equipment, it was observed that the results of the MSA showed less variation, as shown in Figure 5.14.

5.6.5   EXAMPLE 3: AI-ASSISTED COSMETIC VISUAL INSPECTIONS VERSUS HUMAN-ONLY INSPECTION

While manufacturing products, an operator must follow standards established for the norms of what is considered the final product with acceptable quality. Only products that pass such quality tests are destined for sale to the consumer. In Brazil, the technical standard that indicates this is NBR 5426 to obtain measurements, tests and/or comparisons to check the list of possible defects according to their severity and identification in what is called the acceptable quality level (AQL) (ABNT, 2005). While manufacturing items with emphasis on the cosmetic appearance of the finished product in the eyes of the final consumer, there are internal processes that include visual inspection as a guarantee that the product is in 'acceptable' condition

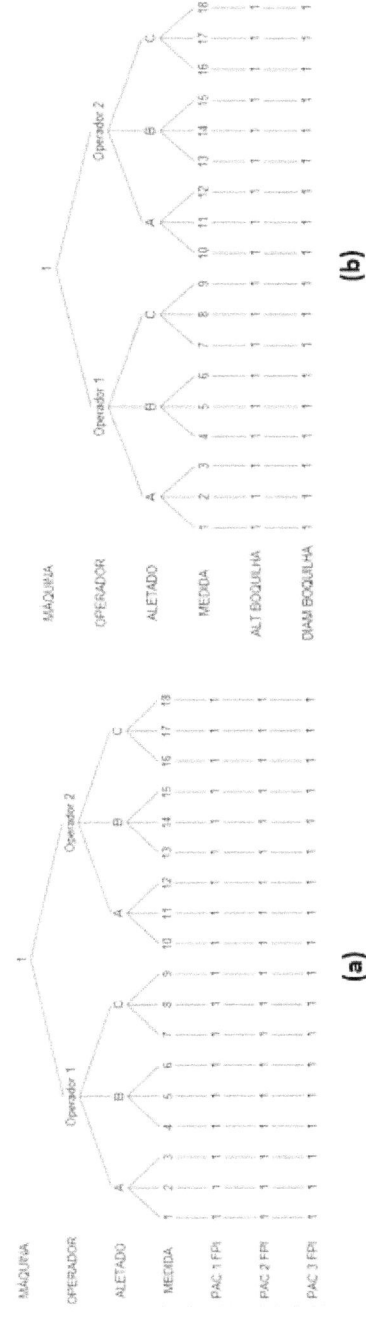

FIGURE 5.12 (a) Sampling trees for FPI data collection; (b) height of mouthpiece diameter.

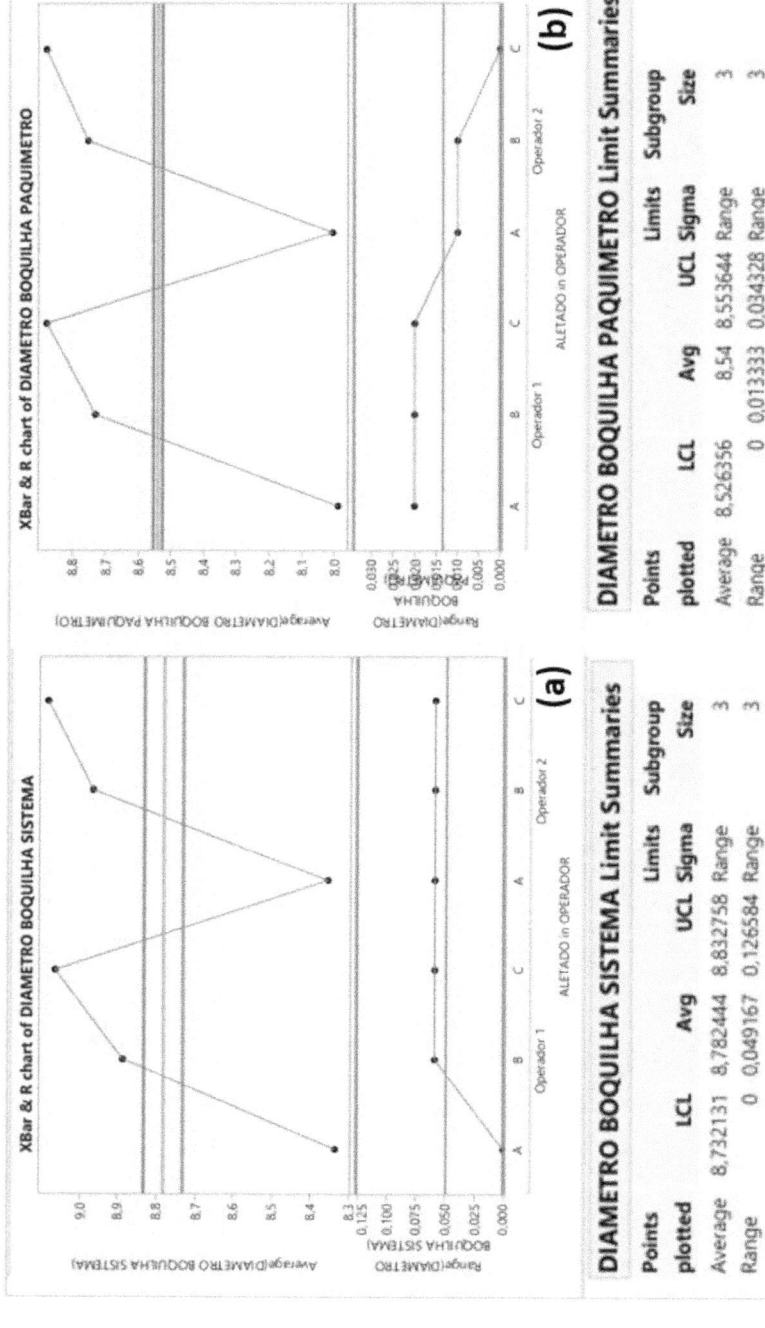

FIGURE 5.13 Comparative MSA before improvements were applied to the system: (a) diameter of the mouthpiece with new system, (b) diameter of the mouthpiece with caliper, (c) height of the mouthpiece with new system, and (d) height of the mouthpiece with caliper.

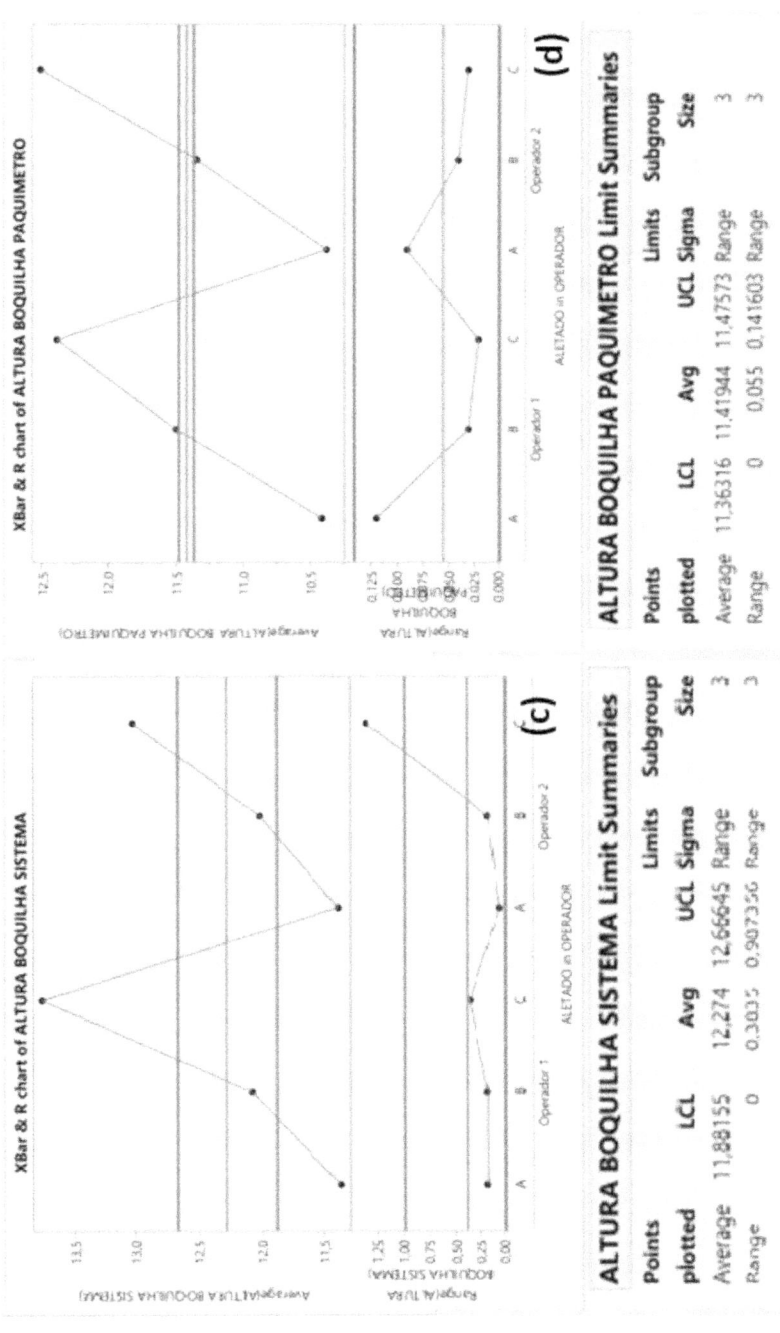

FIGURE 5.13 (Continued)

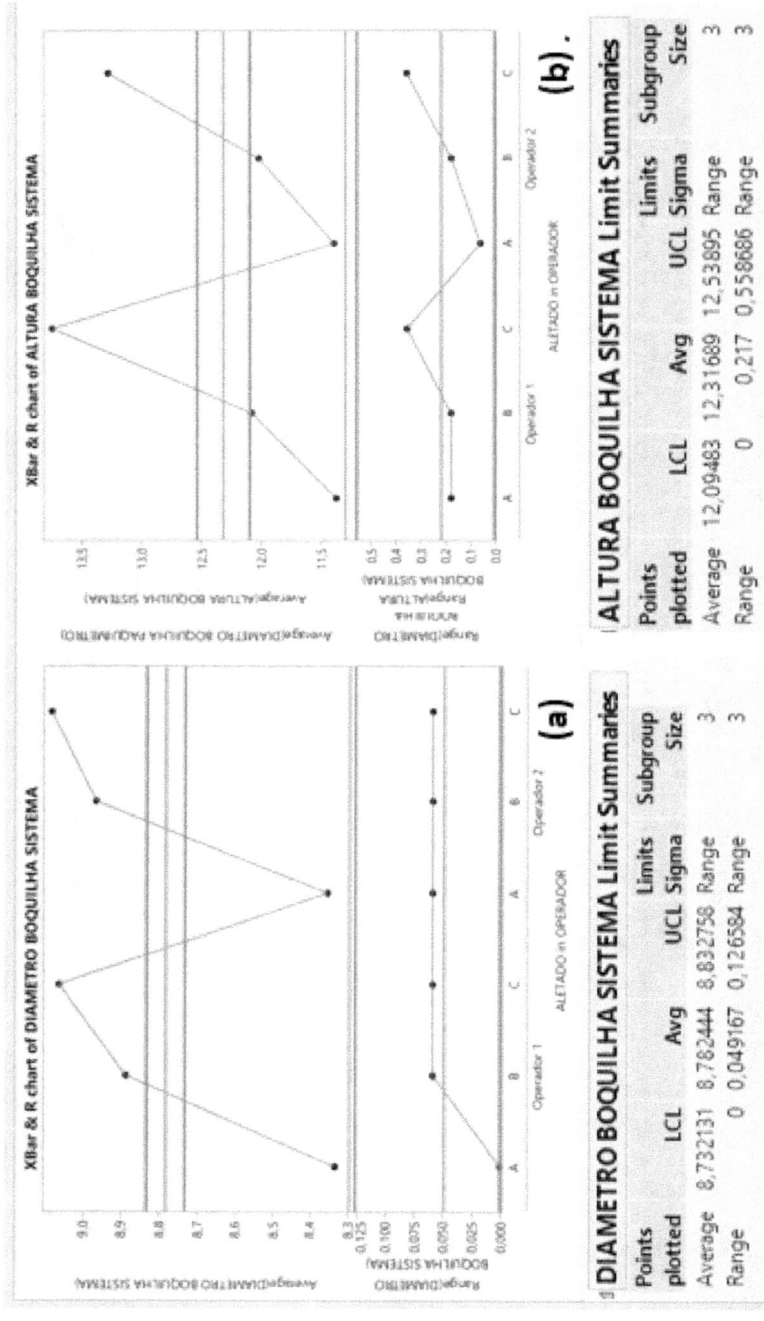

FIGURE 5.14 Comparative MSA after improvements applied to the system: (a) nozzle diameter with new system; (b) nozzle height with new system.

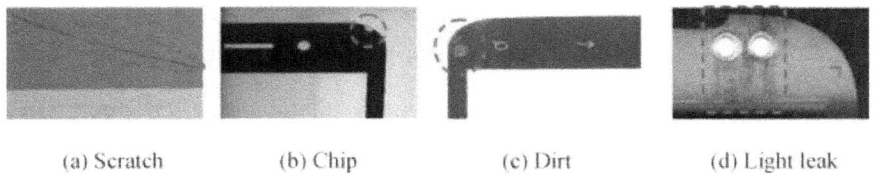

(a) Scratch   (b) Chip   (c) Dirt   (d) Light leak

FIGURE 5.15   Common cosmetic defects as 'areas of interest'.

Source: Jian et al., 2017

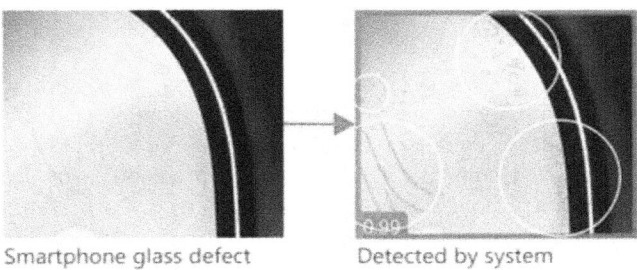

Smartphone glass defect   Detected by system

FIGURE 5.16   Automatic detection of 'areas of interest' with cosmetic defects.

Source: DWFritz Precision Automation, n.d.

to be packaged and directed for sale. Normally the visual inspection that guarantees the quality of the product is performed by human operators/inspectors; however, repetitive and continuous visual inspection can impact the quality of vision by creating eye strain, fatigue and stress in the worker due to a long periodic of such activity (Vilhena et al., 2018).

Therefore, the visual inspection work itself, which guarantees quality, being carried out by a human being, even if well directed by the AQL, is subject to errors at the time of its application. Considering this, and looking at existing technologies, there are currently techniques of computer vision system (CVS). With the technology advancing, methods of intelligent computer systems based on images similar to human vision have emerged capable of being processed through computers, or computer vision. In the context of the presented case, that is, detection of visual cosmetic defects on smartphone devices, there are multiple works that demonstrate various machine learning–based approaches to detect surface defects through computers (Jian et al., 2017; J. Park et al., 2020; J. K. Park et al., 2016). Common cosmetic defects are shown in Figure 5.15.

In Figure 5.16, smartphone glass surface defects, automatically detected by one commercially available product by (DWFritz Precision Automation, n.d.), are shown.

As these works indicate, image capture alone is not capable of detecting deeper defects. For this, it is necessary to perform and simulate human functions such as the ability to learn, adapt and make decisions in different situations, and the aim of AI is to walk this path (Russell & Norvig, 1995). As discussed earlier, there are several approaches and algorithms to simulate the human brain and our logical inference methods. The human brain, despite its tremendous capabilities of thinking and logic, has certain limitations. One such limitation is getting tired or developing fatigue after a long duration of work and repetitive tasks. The hypothesis in this example is that the human worker with his or her high intelligence, assisted by an AI system that does not get fatigued, is an efficient and effective combination, instead of the two working separately. The data presented and discussed here is actual data obtained from a smartphone manufacturing company in the state of Manaus, Brazil.

5.6.6 BUSINESS AND DATA UNDERSTANDING

The general processes of the smartphone company are related according to the macro layout in Figure 5.17. In the image, the cosmetic inspection process is highlighted at Station 5, and in addition, there is a sample at Station 7 for visual inspection in order to review a few selected samples if the smartphones are good cosmetically.

The processes corresponding to each function are given in Table 5.7. During the monitoring of the process, in the period of six months of cosmetic inspection of smartphones (inspection of smartphones' cosmetic quality, whether there is any visible defect in appearance, such as scratches etc.) carried out by human operators

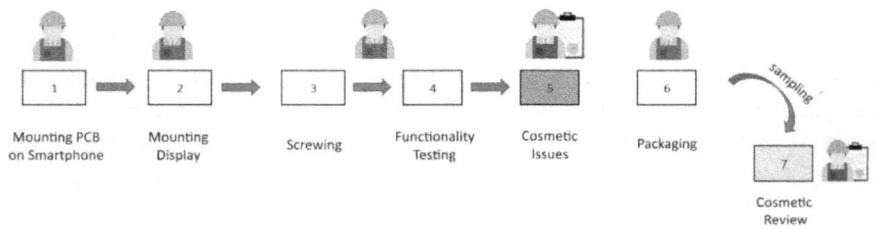

FIGURE 5.17   Macro smartphone assembly process.

TABLE 5.7
Macro Functions of Mounting Smartphones

No	Process	Assembly
1	Mounting PCBS on smartphone	PCBS, battery, cables, labels
2	Mount display	Screen, labels, microphone, camera
3	Screwing	Closing
4	Function tests	Color test, screen, audio, network
5	Cosmetic issues	Check strokes, scratches, impurities
6	Packing	Mount phone and accessories in the box
7	Cosmetic review	Review cosmetic appearance (~5%)

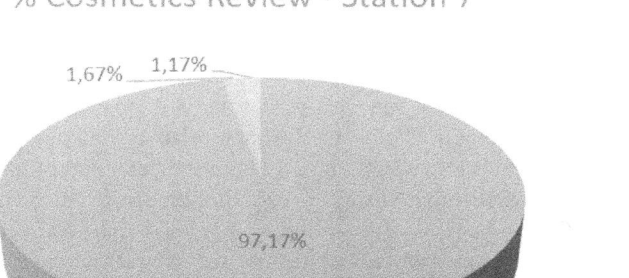

FIGURE 5.18 Percentage of cosmetic inspection—Station 7.

(inspectors), it was observed that there were defective pieces that passed through the operators as if they were 'good' products and, similarly, good products, that is, those without any visible cosmetic defects, were labeled defective by human operators.

The portions of good quality product and cosmetically defective product, out of all the smartphones manufactured in the company and sampled at Station 7, are highlighted in Figure 5.18. This index is based on the behavioral average of monthly production of the company. Looking at the graph in Figure 5.18, which represents inspection performed by human operators, it is evident that there were failures that occurred in the judgment of the inspection station, Station 5. This could have been caused by several factors. Considering the data collected, the biggest challenge is to reduce the 1.67% of false negatives, that is, the undetected 'defective' phones. This could mean that the system may pass some cosmetically defective phones to consumers.

However, in this company, there is another filter that reviews the smartphones again before packaging. Therefore, the main objective of the company was to guarantee that it would not pass cosmetic defects in production, without the need for a second revision before packaging the products. Hence, it installed an automatic cosmetic inspection machine to recognize cosmetic defects by capturing images through computer vision assisted by artificial intelligence using reinforcement learning and neural network (NN) algorithms. In this case, the model is trained with an image bank of cosmetic defects. However, the company had not yet made an efficiency comparison to see whether the station would be meeting its needs. Because of this, a case was made comparing the efficiencies of both the AI station and the human operators, which we will see in the following.

5.6.7 System Validation

During the 20 days of smartphone production in the factory, a total of 1824 samples were collected for the present study, produced in three batches, to compare the efficiency of the operator and the AI station. The objective was to understand the effectiveness of using artificial intelligence in the day-to-day processes in the production facility in order to help employees in the detection of cosmetic defects. The smartphones passed at the AI station were the same as those evaluated by the human operators, without one knowing what the other had identified. Again, since the objective of the inspection was to detect defects, the samples with defects were labeled as 'positives' and the others 'negatives'.

After the period of collection and evaluation by both the operator and the AI station, the results are compiled in Table 5.8.

In order to evaluate and compare the two inspection methods, the values from the confusion matrix for each method from Table 5.8 are then applied to calculate AI system evaluation metrics as in previous examples. The values of these metrics are presented in Table 5.9.

Based on the results of Table 5.9, it is necessary to understand what the client asked for as a priority in business and data understanding, which was to avoid passing cosmetic defects, and this is represented by what we call recall; that is, the greater the recall, the greater the capacity to identify defects. Considering what has been measured, the AI station has a greater capacity to detect defects as compared to the human operator's station. This helps achieve the objective of this company. Although

TABLE 5.8

Smartphone Inspection Results—Confusion Matrix Data for Both Inspection Methods

	Without Defects		With Defects		Total
	True Negatives	False Positives	True Positives	False Negatives	
Manual Inspection	1557	12	194	61	1824
AI Station	1371	198	255	0	1824

TABLE 5.9

Results of Analysis with AI Parameters with Data from the Operator and the AI Station

	Accuracy	Precision	% False Positives	% False Negatives	Recall
Manual Inspection	96.00%	94.17%	0.66%	3.34%	76.08%
AI Station results	89.14%	56.29%	10.86%	0.00%	100.00%

the operator had greater accuracy, this does not represent the result that the client expects. This may have occurred due to visual stress to the worker due to a long period of this activity, the changing of operators over three shifts or some other issues which are not within the scope of this study.

However, it is clear that the AI-powered station helped achieve the goal under the given circumstances. In other words, after the comparison of the results in terms of evaluation parameters from both methods, it may be declared helpful to have the AI-powered station help employees/inspectors in the detection of cosmetic defects in daily routine of the production process. Other than providing better quality and efficiency, the AI-powered station may help reduce the cost of quality control as well.

5.7   CONCLUSIONS

This chapter aims to present basic concepts in machine learning, showing its historical background and how ML algorithms can be applied and evaluated in practical cases in industry, considering the aspects of sustainability, digital transformation, productivity, quality and costs. To develop a machine learning–based application project, CRISP-DM methodology was presented and applied as a guideline, showing the steps to implement a solution to improve different types of industrial systems. The first example shows how to implement a machine learning project using a toolbox, in this case Orange, and test multiple algorithms for the same problem. The second example shows how a machine learning–based system could help improve statistical quality control. Finally, the third example demonstrates how machine learning–assisted decision-making helps improve the quality of an otherwise monotonous industrial task that requires constant inspection, causing fatigue in human inspectors.

For the first example, only a few samples were available for training and testing. It is expected that with more data for training, the system could be better trained and will deliver better results. A bigger sample also allows more robust testing of the hypothesis. On the other hand, the metrology application shows that statistical data based on Six Sigma methodologies can also be used to check algorithm performance compared to other measurement systems, and measurement error can be precisely calculated, serving as a decision factor for system evaluation. The main objective for the chapter was achieved by showing the application of machine learning in some real-world industrial applications and their benefits for everyday manufacturing operations in terms of efficiency, quality and costs. These technologies make the current manufacturing industry more efficient and continuously competitive, for example, with better quality control, bigger production volumes, reducing human stress from repetitive labor (with occupational risks and avoiding human-induced errors) and gaining more accuracy through artificial intelligence–powered machines with human support, thus achieving better results. Contrary to the third industrial revolution, which was a digital revolution, where computers started replacing humans for many 'intelligent' tasks, the current revolution has 'humans' in the centre, despite some of the 'intelligence-intensive', 'smart' and 'decision-making' tasks being passed on to the machines, as emphasized in the two part publication by Putnik et al. (2020, 2021a, 2021b). Thus, the material presented in this chapter supports the Industry 4.0 vision.

ACKNOWLEDGMENTS

This work has been supported by FCT—Fundação para a Ciência e Tecnologia within the R&D Units Project Scope: UIDB/00319/2020 and EXPL/EME-SIS/1224/2021.

REFERENCES

ABNT. (2005). *NBR 5426*. Planos de amostragem e procedimentos na inspeção por atributos.

Angelopoulos, A., Michailidis, E. T., Nomikos, N., Trakadas, P., Hatziefremidis, A., Voliotis, S., & Zahariadis, T. (2020). Tackling faults in the Industry 4.0 era—a survey of machine-learning solutions and key aspects. *Sensors (Switzerland)*, 20(1). MDPI AG.

Camilo, C. O., & Silva, J. C. da. (2009). *Mineração de Dados: Conceitos, Tarefas, Métodos e Ferramentas*. https://rozero.webcindario.com/disciplinas/fbmg/dm/RT-INF_001-09.pdf

Côrtes, S. D. C., Porcaro, R. M., & Lifschitz, S. (2002). Mineração de Dados—Funcionalidades, Técnicas e Abordagens. *PUC-Rio Informática*, 35.

Data Science Academy. (2019). O Que é Aprendizagem Por Reforço? In *Deep Learning Book*. https://www.deeplearningbook.com.br/o-que-e-aprendizagem-por-reforco/

Data Science Project Management. (1999). *CRISP-DM*. https://www.datascience-pm.com/crisp-dm-2/

Davies, R. (2015). *Industry 4.0 Digitalisation for Productivity and Growth*. https://www.europarl.europa.eu/RegData/etudes/BRIE/2015/568337/EPRS_BRI%282015%29568337_EN.pdf

Demšar, J., Curk, T., Erjavec, A., Gorup, Č., Hočevar, T., Milutinovič, M., . . . & Zupan, B. (2013). Orange: Data Mining Toolbox in Python. *Journal of Machine Learning Research*, 14(1), 2349–2353.

DWFritz Precision Automation. (n.d.). *Smart Cosmetic Defect Detection Increases Productivity | DWFritz Automation*. Retrieved June 14, 2021, from https://www.dwfritz.com/2019/06/30/smart-cosmetic-defect-detection-increases-productivity/

Gonçalves, D. B. (2019). *Machine learning in analytical chemistry: Applying innovative data analysis methods using chromatographic techniques* (Issue October). Universidade do Minho.

Jian, C., Gao, J., & Ao, Y. (2017). Automatic surface defect detection for mobile phone screen glass based on machine vision. *Applied Soft Computing Journal*, 52, 348–358.

Kohavi, R., & Provost, F. (1998). Glossary of Terms. *Scandinavian Journal of Public Health*, 271–274.

Lin, M. (2018). *Data Science Cheatsheet*. http://mavericklin.com

Manupati, V. K., Akhtar, M. D., Varela, M. L. R., Putnik, G. D., Trojanowska, J., & Machado, J. (2018, March). A text mining based supervised learning algorithm for classification of manufacturing suppliers. In *World Conference on Information Systems and Technologies*, 236–244, Springer.

Matos, P. F., Lombardi, L. de O., Ciferri, R. R., Pardo, T. A. S., Ciferri, C. D. de A., & Vieira, M. T. P. (2009). *Relatório Técnico "Métricas de Avaliação."* http://conteudo.icmc.usp.br/pessoas/taspardo/TechReportUFSCar2009a-MatosEtAl.pdf

Mohammed, M., Khan, M. B., & Bashier, E. B. M. (2016). Machine learning, algorithms and applications. *Journal of Chemical Information and Modeling*, 53(9). CRC Press.

Ohde, C., & Mattar, H. (2018). *Circular Economy: A Model that Gives Impetus to the Economy, Creates Jobs and Protects the Environment*. Netpress Books.

Park, J., Riaz, H., Kim, H., & Kim, J. (2020). Advanced cover glass defect detection and classification based on multi-DNN model. *Manufacturing Letters*, 23, 53–61. https://doi.org/10.1016/j.mfglet.2019.12.006

Park, J. K., Kwon, B. K., Park, J. H., & Kang, D. J. (2016). Machine learning-based imaging system for surface defect inspection. *International Journal of Precision Engineering and Manufacturing—Green Technology*, 3(3), 303–310.

Putnik, G. D., Shah, V., Putnik, Z., & Ferreira, L. (2020). Machine learning in cyber-physical systems and manufacturing singularity—it does not mean total automation, human is still in the centre: Part I—manufacturing singularity and an intelligent machine architecture. *Journal of Machine Engineering*, 20(4), 161–184.

Putnik, G. D., Shah, V., Putnik, Z., & Ferreira, L. (2021a). Machine learning in cyber-physical systems and manufacturing singularity—it does not mean total automation, human is still in the centre: Part II—in-CPS and a view from community on Industry 4.0 impact on society. *Journal of Machine Engineering*, 21(1), 133–153.

Putnik, G. D., Pabba, S. K., Manupati, V. K., Varela, M. L. R., & Ferreira, F. (2021b). Semi-double-loop machine learning based CPS approach for predictive maintenance in manufacturing system based on machine status indications. *CIRP Annals—Manufacturing Technology*, 70(1), 365–368.

Ribeiro, A. C. F., Frazão, R., & Oliveira e Sá, J. (2018). Machine learning puzzles: How to select use cases, algorithms and technologies? *18ᵃ Conferência Da Associação Portuguesa de Sistemas de Informação (2018)*. AISNET.

Ribeiro, T. A. (2019). *Deep Reinforcement Learning for Robot Navigation Systems* [Universidade do Minho]. http://repositorium.sdum.uminho.pt/handle/1822/64866

Roh, Y., Heo, G., & Euijong Whang, S. (2019). A Survey on Data Collection for Machine Learning a Big Data-AI Integration Perspective. *IEEE Transactions on Knowledge and Data Engineering*, 33(4), 1328–1347.

Russell, S. J., & Norvig, P. (1995). Artificial intelligence A modern approach. In *Pearson*. PearsonEducation.

Rüßmann, M., Lorenz, M., Gerbert, P., Waldner, M., Engel, P., Harnisch, M., & Justus, J. (2015). *Industry 4.0: The Future of Productivity and Growth in Manufacturing Industries*. Boston Consulting Group. https://www.bcg.com/publications/2015/engineered_products_project_business_industry_4_future_productivity_growth_manufacturing_industries

Saltz, J. (2020). *CRISP-DM Is Still the Most Popular Framework for Executing Data Science Projects—Data Science Project Management*. https://www.datascience-pm.com/crisp-dm-still-most-popular/

Samuel, A. L. (1959). *Some Studies in Machine Learning Using the Game of Checkers*. IBM.

Shah, V., & Putnik, G. D. (2019). Machine learning based manufacturing control system for intelligent cyber-physical systems. *FME Transactions*, 47(4), 802–809.

Shearer, C. (2000). The CRISP-DM model: The new blueprint for data mining. *Journal, 5*. www.dw-institute.com/journal.htm

Turing, A. M. (1950). Computing machinery and intelligence. *Mind, LIX*(236), 433–460.

Vilhena, D., Freitas, S., Guimarães, M., & Pinheiro, A. (2018). *The Role of the Psych Pedagogue in the Identification and Intervention in Learning Disorders Related to Vision: Case of a Late Intervention*. Paidéia.

Wuest, T., Weimer, D., Irgens, C., & Thoben, K.-D. (2016). Machine learning in manufacturing: Advantages, challenges, and applications. *Production & Manufacturing Research*, 4(1), 23–45.

Xu, H., Yu, W., Griffith, D., & Golmie, N. (2018). A survey on Industrial Internet of Things: A cyber-physical systems perspective. In *IEEE Access* (Vol. 6, pp. 78238–78259). Institute of Electrical and Electronics Engineers Inc.

6 Evaluating the Make-to-Order Performance of Production Control Systems in a Re-Entrant Flow Shop

Ana Araújo, Margarida Pires, Maria Pereira, Sónia Ribeiro, Maria Leonilde Rocha Varela, Marcelo Henriques, Luís Dias, Guilherme Pereira, Nuno O. Fernandes and Sílvio Carmo-Silva

CONTENTS

6.1 INTRODUCTION

In today's marketplace, companies compete on the ability to deliver their products on time, as this is fundamental in the scope of a sustainable Industry 4.0 (Varela et al., 2019). Clients are getting more attentive and giving more importance to customizability in the customization era (Castro et al., 2010; Zawadzki & Żywicki, 2016), along with the need for flexible manufacturing environments, such as flexible flow shops (Varela et al., 2017) or job shops (Reddy et al., 2017) and, further, re-entrant flow shops (Dugardin et al., 2010), which is the configuration focused on in this chapter. This manufacturing system configuration can further be put to work

DOI: 10.1201/9781003123866-6

119

in a company as a centralized manufacturing environment or through an extended or distributed manufacturing environment (Varela et al., 2012, Varela & Ribeiro, 2014; Arrais-Castro et al., 2018). In these manufacturing environments—intended to closely support and further collaborate with clients, as well as with other stakeholders—in the context of networked manufacturing, such as suppliers and business partners, production control systems for make-to-order clearly influence this ability (Jassbi et al., 2014; Rahman et al., 2021). Thus, alongside companies' internal performance objective satisfaction, external considerations—more related to client requirements and satisfaction—also have to be considered (Arrais-Castro et al., 2018), as this plays a crucial role, nowadays, in enabling sustainable manufacturing and management in Industry 4.0 (Varela et al., 2019). Several production control systems have been proposed in the literature, each with its own way to control the release and flow of jobs in the production system, and, currently, in the scope of Industry 4.0, new insights are being given further in the direction of the exploration of autonomous production control strategies and approaches (Martins et al., 2020a, 2020b).

In general, either with or without autonomous strategies, by controlling the jobs that are allowed to initiate production, it is possible to control the level of work in progress (WIP) on the shop floor and thus job throughput times (Wiendahl, 1992), as it is an important company internal performance measure to be considered in production activity control (Gajsek et al., 2019). Two well-known production control systems for make-to-order are workload control (WLC) (Zäpfel & Missbauer, 1993; Stevenson et al., 2005) and the generic Kanban system (GKS) (Chang & Yih, 1994). These implement different approaches to job release. While the former defines workload norms at critical workstations and releases jobs, taking into consideration the resulting workload at these stations, the latter uses Kanban cards, which are attached to the jobs at release, implementing a simple but effective approach. This chapter examines the behavior of these production control systems in the context of a re-entrant flow shop.

While there is a broad literature on re-entrant flow shops, only recent studies (e.g., Thürer and Stevenson, 2016; Neuner & Haeussler, 2021) have highlighted the potential impact of WLC on the performance of such systems, where the re-entrant nature of routing leads to workstations being visited one or more times by jobs. Thus, in this chapter, through the use of discrete event simulation, we seek to examine the performance of a card-based approach, GKS, comparing its performance with WLC in this context. The remainder of the chapter is structured as follows. Section 6.2 outlines the different production control systems considered in the study. Section 6.3 then presents the method used to evaluate performance, before results are presented and discussed in Section 6.4. Finally, conclusions and directions for future research work are summarized in Section 6.5.

6.2 BACKGROUND

This section introduces the production control systems considered in this study, WLC and GKS. For a literature review on the WLC production control system, we refer the reader to, for example, Thürer et al. (2011).

6.2.1 Workload Control

WLC (Zäpfel & Missbauer, 1993; Land & Gaalman, 1996; Stevenson et al., 2005; Henrich et al., 2004) is a production control system based on the principle of input-output control, which is particularly suited for make-to-order. An important decision of WLC is job release. Job release regulates the level of work-in-process on the shop floor, restricting and balancing workloads at production stages. This assumes that arriving jobs wait for a release decision in a pre-shop pool. This decision is usually based on the urgency of the job and on the current shop floor situation, that is, primarily on workload.

WLC essentially controls the release of job to the shop floor based on a limit on the workload allowed in every production stage, usually referred to as a workload norm. Jobs are released if they do not overload any production stage; that is, the calculated workload on every production stage resulting from the release of the job is not larger than workload norm; otherwise, the job is kept in the pre-shop pool. The job release order is according to a job ordering process based on a chosen priority rule, referred to as a sequencing rule. Every time the job releasing process starts, it ends after all jobs waiting to be released are considered for release. The release process starts either every time a job operation is finished or periodically according to a time period of release. In the former situation, the release is said to be continuous, and in the latter, it is periodic. The calculated workload includes the load of the operations to be processed or that are unfinished of the jobs in the shop and that of the operations of the job considered for release. It has been concluded that this calculation should be based on the corrected aggregate load method (Oosterman et al., 2000). This method accounts only for the load that results from dividing the workload of each of the job's operations by the position of the production stage, which processes the operation in the routing of the job.

In the study presented in this chapter, the corrected aggregate method is used, and job release is continuous.

6.2.2 Generic Kanban System

GKS (Chang & Yih, 1994) is a card-based production control system for make-to-order. GKS provides a fixed number of cards, Kanbans, at each production stage as a means of controlling the work-in-process and output of the production system. These cards are not specific to any product and can be allocated to any job waiting to be released onto the shop floor. A job cannot be released unless it gets all the cards that it needs, one per stage, from all the production stages in its routing. After release, jobs are pushed through the production system, and, as they are completed at each station, the attached cards are detached, becoming available to be allocated to new jobs, which are waiting for release in the pre-shop pool. Since cards control the number of jobs on the shop floor, the number of cards at each station is an important parameter of this production control system. To deal with the re-entrant nature of the routing of jobs, we have adapted GKS to provide a card per operation. This means that, at job release, one or more cards from each production stage, depending on the number of return visits in the routing of the job, are attached to the job.

6.3 METHOD

To compare the performance of the two production control systems under study, a simulation study was carried out using the SIMIO software. This section describes the simulation model built and used, the control policies implemented and the experimental setup considered.

6.3.1 GENERIC KANBAN SYSTEM

A re-entrant flow shop with three production stages is considered. At each production stage, there are two workstations preceded by a single queue. Operations 1 and 5 of all products are carried out at stage 1, operations 2 and 4 are carried out at stage 2 and operations 3 and 6 are carried out at stage 3. Workstations are subject to random failures and available 99% of the time. Operation processing times are stochastic and follow a lognormal distribution with a mean of 1 hour. This distribution has been used in previous simulation studies (e.g., Thürer et al., 2016; Fernandes et al., 2017). Job setup times are sequence independent and considered part of processing times. The inter-arrival time of jobs to the system follows an exponential distribution with a mean of 1.111 hours. This was chosen to ensure that, on average, workstation utilization is 90%. Job due-date allowances are set exogenously by the customer, following an uniform distribution between 15 and 30 hours. The upper and lower limits of the random allowances were set based on the preliminary simulation experiments to result in about 16% of tardy jobs under unrestrictive order release. Material supply to workstations is not considered a restriction to the flow of work. Data of jobs about processing operations, including times and routings, is known when jobs arrive. These two assumptions are common on studies of production control, for example, Thürer et al. (2012) and Fernandes et al. (2017).

Table 6.1 summarizes the operational settings.

6.3.2 PRODUCTION CONTROL

Upon arriving, jobs flow to the backlog or pool that precedes the shop floor and remain there until they are released onto the shop floor. Under WLC, a job is released

---

TABLE 6.1
Summary of the Operational Settings of the Simulation Study

Shop type	Re-entrant flow shop
Production environment	Make-to-order
Number of production stages	3
Number of machines per stage	2
Machine availability Machine utilization rate	99% 90%
Operation processing times	Lognormal with a mean of 1 hour
Number of operations per job	6
Job inter-arrival time	Exponential with a mean of 1.111 hours
Job due date allowance	Uniform [15, 30] hours

for production if it does not violate the workload norms of every production stage in the routing of the job. Under GKS, a job is released if cards from all production stages in the routing of that job are available to be allocated to it. At each production stage, seven workload norms (4, 5, 6.3, 7.8, 12.2, 15.3 and infinity) have been defined for WLC, and seven card counts (15, 19, 23, 29, 37, 46, 57 and infinity) have been defined for GKS. Increments of 25% in the workload norms and card counts have been considered. Infinity, in both cases, means unrestrictive release of jobs to the shop floor and was used as a benchmark in the study.

Jobs in the pre-shop pool are sequenced and considered for release according to the earliest due date (EDD) rule, which considers the urgency of the jobs. Once released, jobs are then controlled on the shop floor through dispatching rules. Therefore, when a workstation becomes idle, a dispatching rule is used to select the job to be processed next. The dispatching rules considered in the study are: (i) first come first served (FCFS) and (ii) the earliest operation due date (ODD), which is obtained by backwards scheduling from the jobs' due date using production stage lead times.

Comparing the behavior of the production control systems, both figures show that, over a large range of the workload norm, WLC performs better than GKS, although under FCFS dispatching, they exhibit similar values of tardiness. However, as long as the number of cards for GKS or the workload norm for WLC is duly tuned, both control systems manage to achieve very similar best delivery performance, measured by the combination of both tardiness and tardy jobs. The performance of the dispatching rules can be observed by comparing Figures 6.1 and 6.2. As can be seen, FCFS performs better than ODD, particularly when GKS is applied. However, for loose values of WIP or workload, the efficiency of FCFS is very poor. Again, this is avoided for duly tuned values of number of cards for GKS or workload norm for WLC, which leads to identical best delivery performance of both GKS and WLC. Thus, the results presented by both figures lead us to conclude that, taking into account the simplicity of using both the FCFS rule and the GKS system, as against ODD and WLC, the best operating combination—as far as the managerial point of view is concerned—is effectively to use GKS with FCFS dispatching.

6.4 CONCLUSIONS

This chapter puts forward a contribution regarding the evaluation of the make-to-order performance of production control systems in a re-entrant flow shop configuration as an important approach to fulfill requirements underlying sustainability issues in the current Industry 4.0 scenario by considering internal companies' performance measures alongside client-oriented ones. Two make-to-order production control systems, workload control and generic Kanban systems, were tested and compared in the context of a re-entrant flow shop under two dispatching rules, FCFS and ODD, using discrete event simulation. The performance of these systems was evaluated for delivery, measured mainly by the percentage of tardy jobs and tardiness and complementarily by the total throughput times of jobs. Our results show that WLC outperforms GKS. However, both systems, when operating under duly tuned card numbers for GKS and workload norms for WLC, managed to achieve identical best levels of performance, which were obtained under FCFS. Thus, taking into account

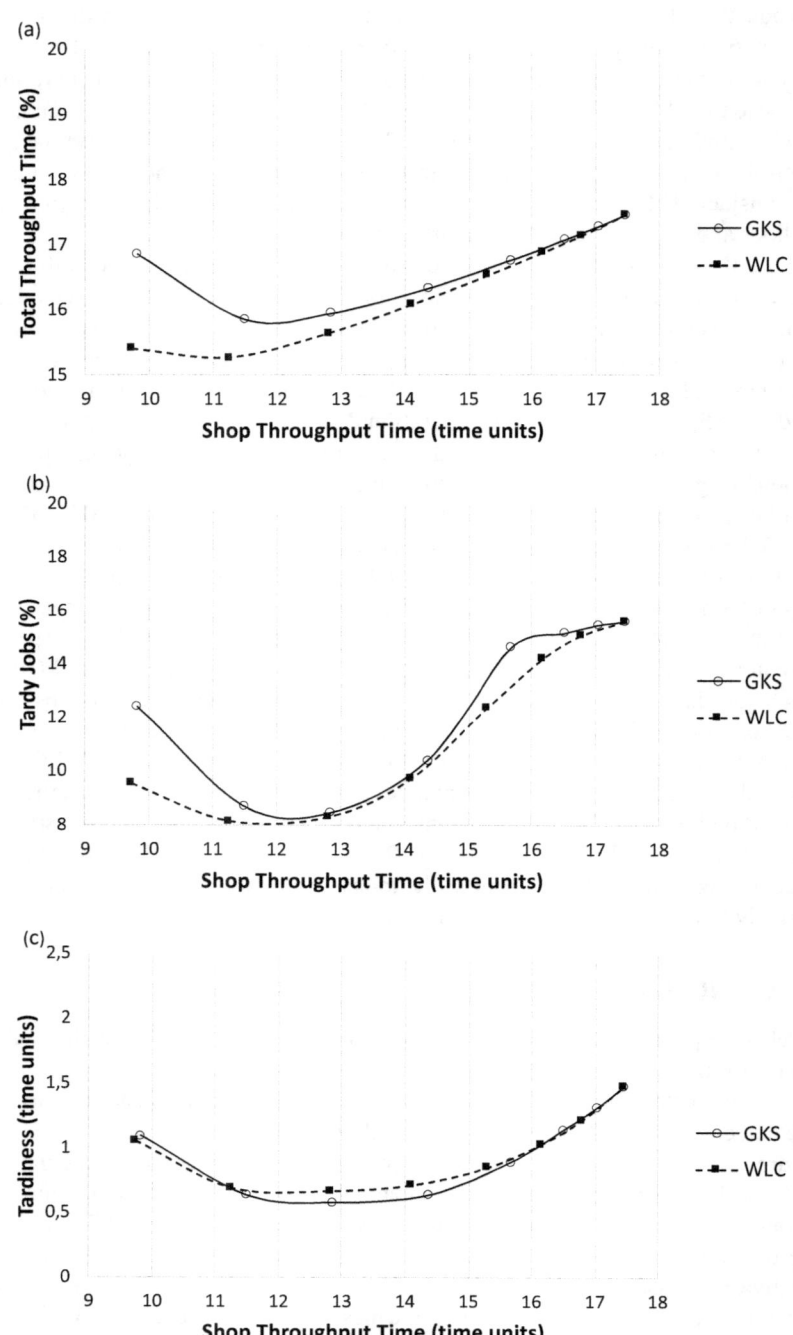

FIGURE 6.1   Performance results under FCFS dispatching.

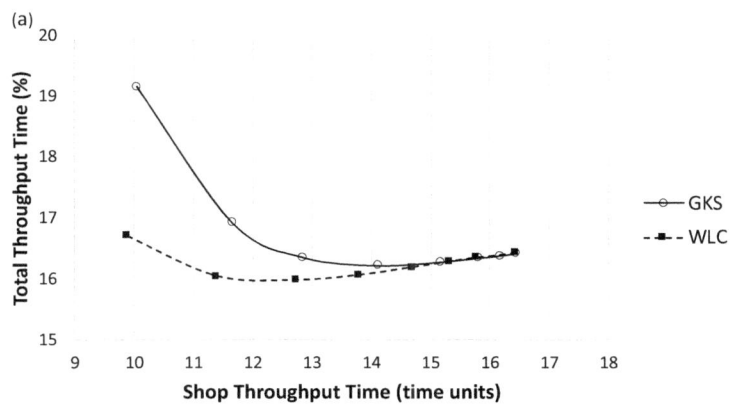

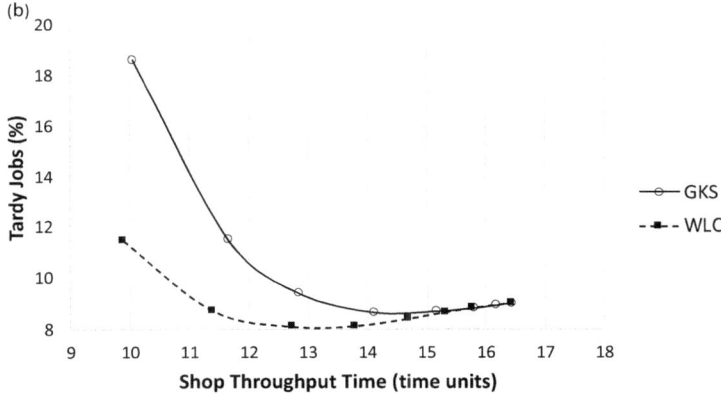

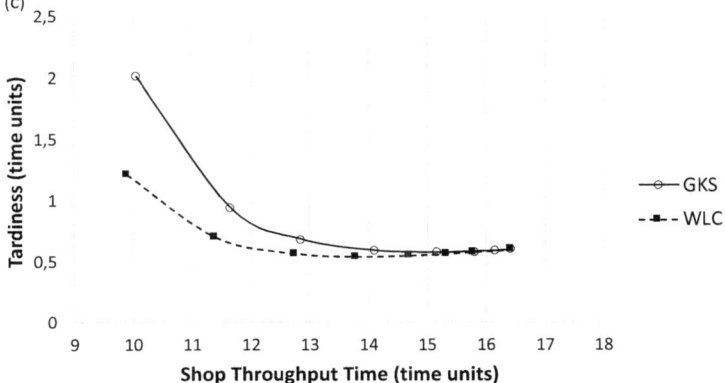

FIGURE 6.2 Performance results under ODD dispatching.

the simplicity of GKS and the FCFS dispatching rule—when compared to the other production control system and dispatching rule—we would recommend the use in practice of this combination for effective and efficient production control under a production environment similar to the one tested, that is, make-to-order of a re-entrant flow line.

Our study was carried out under a somewhat limited experimental setting involving job dispatching and controlled release. Thus, aiming at enlarging the generalization of results, we plan to extend the study to other pool sequencing and dispatching rules, as well as to different card-based production control systems. This would also give a deeper insight into the influence of these factors on the performance of balanced re-entrant flow lines.

ACKNOWLEDGMENTS

This work has been supported by FCT—Fundação para a Ciência e Tecnologia within the R&D Units Project Scope: UIDB/00319/2020 and EXPL/EME-SIS/12 24/2021.

REFERENCES

Arrais-Castro, A., Varela, M. L. R., Putnik, G. D., Ribeiro, R. A., Machado, J., & Ferreira, L. (2018). Collaborative framework for virtual organisation synthesis based on a dynamic multi-criteria decision model. *International Journal of Computer Integrated Manufacturing*, 1–12. Taylor & Francis.

Castro, A. A., Varela, M. L. R., & Carmo-Silva, S. (2010). An architecture for a web service based product configuration information system. *Communications in Computer and Information Science*, 110(PART 2), 20–31. DOI: 10.1007/978-3-642-16419-4_3. (https://link.springer.com/chapter/10.1007/978-3-642-16419-4_3).

Chang, T.-M., & Yih, Y. (1994). Generic Kanban systems for dynamic environments. *International Journal of Production Research*, 32(4), 889–902.

Dugardin, F., Yalaoui, F., & Amodeo, L. (2010). New multi-objective method to solve reentrant hybrid flow shop scheduling problem. *European Journal of Operational Research*, 203(1), 22–31.

Fernandes, N. O., Thurer, M., Silva, C., & Carmo-Silva, S. (2017). Improving workload control order release: Incorporating a starvation avoidance trigger into continuous release. *International Journal of Production Economics*, 194, 181–189.

Gajsek, B., Marolt, J., Rupnik, B., Lerher, T., & Sternad, M. (2019). Using maturity model and discrete-event simulation for Industry 4.0 implementation. *International Journal of Simulation Modelling*, 18(3), 488–499.

Henrich, P., Land, M. J., & Gaalman, G. (2004). Exploring applicability of the workload control concept. *International Journal of Production Economics,* 90(2), 187–198.

Jassbi, J. J., Ribeiro, R. A., & Varela, M. L. R. (2014). Dynamic MCDM with future knowledge for supplier selection. *Journal of Decision Systems, Knowledge-Based Decision Systems*, 23(3), 232–248, Taylor & Francis,

Land, M. J., & Gaalman, G. J. C. (1996). Workload control concepts in job shops: A critical assessment. *International Journal of Production Economics*, 46–47, 535–538.

Martins, L., Fernandes, N., Varela, M. L. R., Dias, L., Pereira, G., & Carmo-Silva, S. (2020a). Comparative study of autonomous production control methods using simulation. *Simulation Modelling Practice and Theory*, 104, 102142, Elsevier, Netherlands.

Martins, L., Varela, M. L. R., Fernandes, N. O., Carmo-Silva, S., & Machado, J. (2020b). Literature review on autonomous production control methods. *Enterprise Information Systems, Taylor & Francis, England*, 14(8), 1219–1231.

Neuner, P., & Haeussler, S. (2021), Rule based workload control in semiconductor manufacturing revisited. *International Journal of Production Research*, (in print)

Oosterman, B., Land, M. J., & Gaalman, G. (2000). The influence of shop characteristics on workload control. *International Journal of Production Economics*, 68(1), 107–119.

Rahman, H. F., Janardhanan, M. N., & Nielsen, P. A. (2021). Smart make-to-order production in a flow shop environment for Industry 4.0. In *Research Anthology on Cross-Industry Challenges of Industry 4.0* (pp. 955–982). IGI Global.

Reddy, M. S., Ratnam, C., Agrawal, R., Varela, M. L. R., Sharma, I., & Manupati, V. K. (2017). Investigation of reconfiguration effect on makespan with social network method for flexible job shop scheduling problem. *Computers & Industrial Engineering, UK,* 110, 231–241.

Stevenson, M., Hendry, L. C., & Kingsman, B. G. (2005). A review of production planning and control: The applicability of key concepts to the make to order industry. *International Journal of Production Research*, 43(5), 869–898.

Thürer, M., & Stevenson, M. (2016). Workload control in job shops with re-entrant flows: An assessment by simulation. *International Journal of Production Research*, 54(17), 5136–5150.

Thürer, M., Stevenson, M., & Silva, C. (2011). Three decades of workload control research: A systematic review of the literature. *International Journal of Production Research*, 49(23), 6905–6935.

Thürer, M., Stevenson, M., Silva, C., Land, M. J., & Fredendall, L. D. (2012). Workload control (WLC) and order release: A lean solution for make-to-order companies. *Production & Operations Management,* 21(5), 939–953.

Varela, L., Araújo, A., Ávila, P., Castro, H., & Putnik, G. (2019). Evaluation of the relation between lean manufacturing, Industry 4.0, and sustainability. *Sustainability*, 11(5), 1439. MDPI.

Varela, M. L. R., Putnik, G. D., & Cruz-Cunha, M. M. (2012). Web-based technologies integration for distributed manufacturing scheduling in a virtual enterprise. *International Journal of Web Portals*, 4(2), 19–34,

Varela, M. L. R., & Ribeiro, R. A. (2014). Distributed manufacturing scheduling based on a dynamic multi-criteria decision model. recent developments and new directions in soft computing. In *Studies in Fuzziness and Soft Computing* (Vol. 317, pp. 81–93). Springer International Publishing, Switzerland. (ISBN: 978-3-319-06322-5).

Varela, M. L. R., Trojanowska, J., Carmo-Silva, S., Costa, N. M. L., & Machado, J. (2017). Comparative simulation study of production scheduling in the hybrid and the parallel flow. *Management and Production Engineering Review*, 8(2), 69–80. DOI: 10.1515/mper-2017-0019.

Wiendahl, H. P., Gläßner, J., & Petermann, D. (1992). Application of load-oriented manufacturing control in industry. *Production Planning & Control*, 3(2), 118–129.

Zäpfel, G., & Missbauer, H. (1993). New concepts for production planning and control. *European Journal of Operational Research*, 67, 297–320.

Zawadzki, P., & Żywicki, K. (2016). Smart product design and production control for effective mass customization in the Industry 4.0 concept. *Management and Production Engineering Review*, 7.

7 Conceptual Multi-Agent System Design for Distributed Scheduling Systems

Filipe Alves, Ana Maria A. C. Rocha,
Ana I. Pereira and Paulo Leitão

CONTENTS

7.1 INTRODUCTION

Nowadays, decision-making processes based on planning and scheduling are very important in the manufacturing domain and industry, economics, health and service management. In the ongoing competitive environment, process scheduling has become crucial to survive in an everyday services environment (Michael, 2018). The goal of scheduling is to allocate different resources to tasks over time in order to optimize one or more objectives, for example, to increase the execution speed; reduce the task runtime; and minimize communication delay, cost and emergent or priority problems, among other goals (Baker and Trietsch, 2013; Leung, 2004; Michael, 2018). Scheduling problems are complex and difficult to solve, considering their large number of integrated entities and interactions; therefore, they are classified in the literature as NP-hard (Pinedo et al., 2015). Scheduling is often encountered in practice and has been studied over the years in operations research literature (Cardoen et al., 2010; Leung, 2004; Michael, 2018; Van den Bergh et al., 2013). Furthermore,

DOI: 10.1201/9781003123866-7

the dynamic nature of scheduling environments makes it difficult to obtain optimal solutions. Hence, several approximation methods (i.e., heuristics, meta-heuristics or agent-based approaches) are some of the proposals to deal with scheduling problems in real environments (Chan and Chung, 2013).

A well-prepared schedule can significantly improve production, lead times, utilization and due-date-related measures and improve the profitability of end products or services. Globalization of production systems has made them more reconfigurable, flexible and demand driven. Centralized scheduling approaches have become obsolete since activities and resources are controlled using just a single decision-maker, because they are less responsive to unexpected events, inflexible to emergencies and unable to respond to the dynamic needs of the market (Leitão, 2009). By drawing a line between the current work and our system, our goal follows the determination of development and implements a complex system that demonstrates in practice (deployed in real-world problems) the benefits of distributed scheduling (DS) subject to unexpected events in dynamic environments. However, organizations and companies have moved from centralized to decentralized architecture due to changes in the business world (Behnamian and Ghomi, 2016). In practice, what distinguishes the DS system from centralized models is the use of system decomposition, since its focus on the communication scheme will allow the solving of sub-problems in different domains that can be combined to form final scheduling. For example, the subcontractors of a major company, who may have alternative processing abilities, would typically prepare their schedules independently from one another but maintain consistency with the full schedule. In a highly distributed environment, local decision-makers are competitors with individual objectives. The need to make scheduling decisions in decentralized systems has given rise to a new area, distributed scheduling. Distributed scheduling is an approach in which smaller parts of a scheduling problem are solved by local decision-makers (local entities) who may have conflicting objectives or specific orders and who coordinate their sub-solutions through certain communication mechanisms to achieve the general goals of the system. Hence, the lower level of the organizational hierarchy delegates and solves decision problems locally and decision problems with different entities and/or resources of the system independently. These entities can be multiprocessors in a communication network, different departments of a company, flexible manufacturing systems or, for example, health care routing or allocation belonging to a national health system.

In a distributed way, decision-making may increase system responsiveness, which is very important for scheduling applications, where a short-term decision requires up-to-date information and data to generate feasible and practical schedules. This chapter aims to contribute to the literature in the area of scheduling, especially in the development of intelligent mechanisms that enable innovative and optimized solutions, reorganization and fast responses to changing conditions in dynamic scheduling. More specifically, the following objectives are defined for the design and specification of a conceptual and intelligent approach that allows an increase of scheduling operating in a decentralized mode, allowing maintenance of high levels of optimal solutions through optimization methods and, at the same time, reducing operation costs with fast responses through MAS. With these advanced mechanisms, namely distributed scheduling, prioritization and dynamic and fast response, it will be possible to achieve the desired adaptation and optimization of task elements in order to achieve self-scheduling or rescheduling for reactive or unexpected events with minimal degradation of the quality of solutions. The architectural design will be based on an intelligent control approach based on the swarm system and MAS design principles, combined with optimization methods that provide dynamic adaptation and optimization capabilities to address the issues where each agent is enhanced with self-organizing oriented mechanisms, communication abilities and control capabilities to retrieve, analyze and forecast data in order to obtain the required information to improve the framework and deploy in real environments of scheduling. DS can be seen as allowing creation of schedules by local decision-makers considering all the objectives, restrictions and constraints within the overall system goals. In terms of the research question, the critical decisions in a DS system, it is possible to observe and identify critical design factors, such as the whole system, individual agents and inter-agent relations. According to the general perspective of the system, it will be possible to define how scheduling problems will or can be decomposed, generate independent decision-making, have independent decision-makers assigned to sub-problems and, for example, enable group decision-making. According to the view of individual agents, it will be possible to identify what information prevails in each of the agents, possible trade-offs between objectives and how the solution can be updated to eliminate conflicts.

The chapter is organized as follows: Section 7.2 overviews the contribution on the main topic and provides a literature review. Section 7.3 describes the preliminary MAS system approach as architecture design, and Section 7.4 presents the preliminary results for the validation methodology with some discussion and a critical view. Finally, Section 7.5 presents conclusions and future work ideas.

7.2 LITERATURE REVIEW

Traditionally, classical optimization algorithms are used to solve scheduling problems, normally considering them static and deterministic. However, for many optimization problems of industrial interest, the computation of optimal solutions by analytical methods is almost impossible (Bartz-Beielstein et al., 2004). Thus, the adaptation of ideas from various areas, mainly based on nature, has resulted in the development of optimization methods, such as nature-inspired or evolutionary metaheuristics. As a result, using metaheuristics to attain a near-optimal solution in a reasonably short period is more realistic than using traditional analytical approaches. However, the dynamic and stochastic nature of industrial environments increases the complexity of the problem, demanding more adaptive and efficient handling of planned or unplanned events in real time, such as changing orders, prioritizing orders, resource breakdown, worker illnesses and large repairs. It is also important to consider the complexity associated with the ramp-up phase of complex and highly customized services, which are exceptionally challenging for allocation, planning, scheduling and control. Taking this into consideration, new methods and information and communication technologies (ICT) are required to develop strategies to respond more quickly to unexpected events, creating novel and emerging approaches to the scheduling problem. In this sequence, complex systems need to adapt their processes with the help of flexible and dynamic infrastructures. For that reason, scheduling perspectives take into account the latest research in the domain of artificial intelligence (AI) and software engineering, especially in the areas of MAS, optimization methods and swarm intelligence, which are increasingly designed and specified approaches for distributed ICT (Lezama et al., 2019). Swarm intelligence is a derived concept that can be defined as "the emergent collective intelligence of groups of simple and single entities" (Bonabeau et al., 1999) reflecting the emergent phenomenon that occurs without a predefined plan and not driven by a central entity. With this approach, swarm principles can be used in a distributed scheduling architecture providing a network of schedulers, organized as a swarm, where each one is responsible for the schedule in an organizational unit of an enterprise or specific service, such as a factory, cell or health unit, among others.

The increased complexity of architectures and problems in computer science required the development of parallel and distributed programming to take advantage of new software and hardware architectures. Distributed artificial intelligence (DAI) was born within AI to meet the requirements for the development of complex distributed systems. The MAS paradigm is one example of DAI (Ferber, 1999) composed of an intelligent society and based on entities, especially cooperative and autonomous ones, called agents. These entities coordinate and interact according to their tasks and activities, making use of their knowledge and skills in order to achieve their goals in a global way. This paradigm can provide innovative management, design and modelling for complex, dynamic and heterogeneous distributed systems (Weiss, 1999). In contrast with traditional centralized and hierarchical approaches that split the problem into hierarchically dependent functions, MAS performs its activities in parallel and thus is characterized by its decentralization.

The agent concept does not have a consensus or unique definition, mainly due to its features and attributes. Some proposed definitions that can be found in the literature are:

- "An agent is a persistent computation that can perceive its environment and reason and act both alone and with other agents. The key concepts in this definition are interoperability and autonomy" (Singh, 1998).
- "An agent is a computational entity that can be viewed as perceiving and acting upon its environment, that is autonomous and that operates flexibly and rationally in a variety of environmental circumstance" (Weiss, 1999).
- "An agent is a computer system that is situated in an environment and that is capable of autonomous action in this environment in order to meet its design objectives" (Wooldridge, 2009).

Independently of its definition, an agent is characterized by its ability to act and adapt flexibly, responding effectively to changes in the environment by sensing the environmental changes in states. It also has autonomy and proactivity that allow it to take initiative to meet its own goals (Wooldridge and Jennings, 1995). Thus, normally, an abstract representation of an agent presents it equipped with sensors and actuators, which are responsible for interacting with the environment, as well as with a reasoning mechanism responsible for determining what actions the agent should perform according to its perceptions and internal state.

Agents are characterized by sets of properties that determine their behavior. The following are the main properties that an agent must have (Wooldridge, 2009; Wooldridge and Jennings, 1995):

- Autonomy—operation without human intervention or systems getting control of their actions and internal states.
- Social ability—interactions with other agents to achieve their goals through specific languages and communication protocols, which allow them to negotiate and cooperate rather than simply exchanging information.
- Reactivity—ability to observe the surrounding environment and being empowered to respond as changes in the system arise.
- Proactivity—ability to display goal-driven behaviors, taking the initiative to reach their goals, rather than simply acting in response to environmental stimuli.

A single agent in some cases can be enough to control, monitor or carry out the task. However, some tasks need to be performed by more than one agent due to their size and/or complexity, giving rise to the concept of MAS. Each agent, despite the need to meet its own goals, can act on behalf of different users or perform tasks to meet the purposes of an application. To succeed in their interactions, some capabilities are required for agents (Russell and Norvig, 2002):

- Cooperation—the ability to work among them to achieve common objectives.
- Coordination—the ability to manage the interdependencies between activities, for example, to use non-shareable resources or to perform a sequence of tasks.
- Negotiation—the ability to use communication mechanisms, such as bidding (e.g., Contract-Net protocol [Smith, 1980]). The negotiation mechanism has the ability to make agreements on issues of common interest, for example, by an offer and a counteroffer of the involved parties.

This interaction is supported by the use of a communication language, agent communication language (ACL), and/or by a set of interaction protocols. There are two major ACLs, Knowledge Query and Manipulation Language (KQML) (Finin et al., 1994) and the Foundation for Intelligent Physical Agents based on Agent Communication Language (FIPA-ACL) (Labrou et al., 1999). Both languages use the theory of speech acts to provide semantics to messages through a set of speech acts, called performatives (Wooldridge, 1997). Performatives are used by agents to represent or interpret their requests and intentions; thus, when a message is received, an agent can understand the sender's intentions and decide what to do. For example, some performatives specified by FIPA-ACL are: *Inform*—used to indicate that the agent is communicating information or a fact, *Request*—used to indicate that the agent is requesting a service or information, *Agree*—used to indicate that the agent agrees to a request from another agent and *Not Understood*—used to indicate that the agent did not understand the received message. While the agent communication language specifies the messages of agents, interaction protocols specify the sequence of messages exchanged between two or more agents in a given scenario, such as an information or service request, as well as for negotiation and cooperation.

In recent years, several agent platforms and frameworks have been developed (Bordini et al., 2006; Kravari and Bassiliades, 2015). The purpose of these tools is to simplify the development of agents to support developers with basic agent infrastructure, such as agent communication, negotiation, collaboration and other features inherent to the MAS approach. Generally, an agent platform offers a set of built-in features for the development and execution of agent-based approaches. In order to ensure interoperability between heterogeneous agents developed on different platforms, many standards and specifications for agent technology were developed. In this context, FIPA[1] is characterized by being an organization that specifies and defines a set of standards developed with the intention to promote interoperability between heterogeneous agents. FIPA standards are widely accepted by the community, and there are several platforms and compatible frameworks, such as JADE.[2] There are also frameworks and platforms developed exclusively for agent-based simulation systems, such as NetLogo.[3]

Therefore, MAS has multidisciplinary applications, such as dynamic products (Wang et al., 2016), manufacturing control (Monostori et al., 2006), production planning (Trentesaux et al., 2013), logistics (Jabeur et al., 2017), health (Nealon, 2003) or home health care (Marcon et al., 2017) and many others (Leitão et al., 2012).

7.3   MAS SYSTEM APPROACH

This section aims to explore and specify a conceptual MAS-based architecture that intelligently searches for, selects and composes distributed scheduling solutions.

In this sense, and considering the requirements of scheduling problems, their characteristics and their restrictions, the proposal of a conceptual MAS should take into account all assumptions and, if possible, contain some of the principles of swarm intelligence, which may facilitate the distributed scheduling structure. Swarm principles or guidelines generally refer to solving a problem with an ability that emerges from the interaction of simple information-processing units. The concept of the swarm suggests multiplicity, distribution, stochasticity, randomness and disorder. In this context, the swarm concept is a new computational and behavioral paradigm for solving distributed problems based on self-organization applied, for example, to scheduling. On the other hand, most of the previous scheduling strategies and approaches consider scheduling static, deterministic and normally centralized. However, several domains of scheduling are subject to a dynamic environment, with new jobs continuously arriving to the system, certain resources becoming unavailable, disruptions and additional resources or orders introduced. The swarm approach offers plenty of powerful mechanisms to handle emerging and evolving environments, where complex systems are built upon entities that exhibit simple behaviors and have reduced cognitive abilities (betting on the hypothesis of a swarm approach). In fact, everybody knows that "a single ant or bee isn't smart but their colonies are"

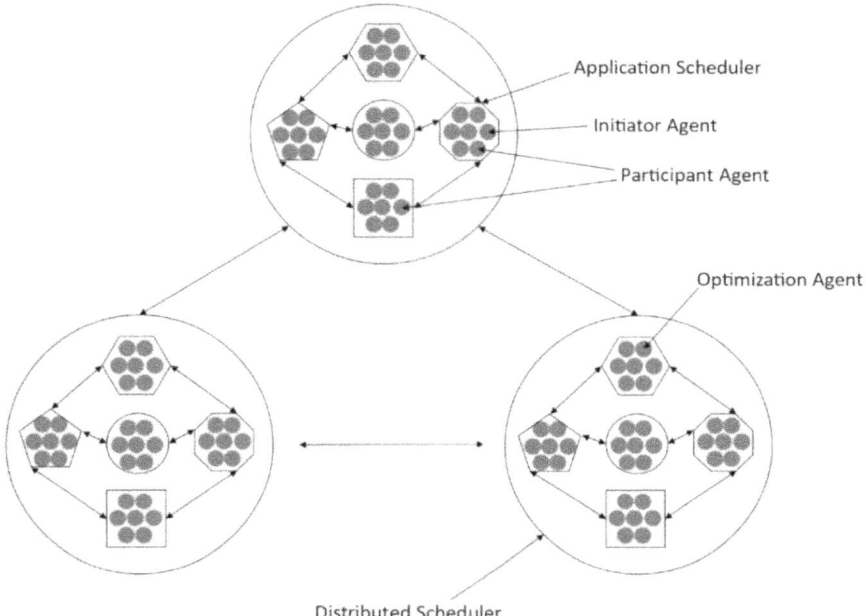

Application Scheduler

Initiator Agent

Participant Agent

Optimization Agent

Distributed Scheduler

FIGURE 7.1   Recursive MAS architecture for a distributed scheduling.

(Miller, 2007) and also that they are capable of exhibiting very surprising complex behaviors. In such environments, the coordination of activities uses simple feedback coordination mechanisms in contrast to traditionally rigid centralized control (betting on the hypothesis of a decentralized approach).

To design a main scheme of the swarm, for a more adaptive and efficient scheduling solution and optimization of intelligence capabilities, it is necessary to design tools for several domains. For this reason, each DS system can be considered a network of schedulers, as shown in Figure 7.1 (general case). Each distributed scheduler is responsible for the schedule of an organizational unit of a service or company, that is, a factory, workshop or station or a health unit belonging to a network of health units.

In this manner, a swarm is a network of individual schedulers, $S = \{S^{(1)}, S^{(2)}, S^{(3)}, \ldots, S^{(n)}\}$, where the global schedule emerges from all the interactions among these schedulers. In this sense, complex systems, like scheduling, are built upon entities that exhibit simple behaviors and have reduced cognitive abilities. For example, in the concept of the swarm, it is found in the literature that simple entities are based on principles of extreme utility, such as the fact that they are regulated by simple rules, have mainly reactive behavior and usually interact with each other and their environment without the existence of a central authority.

The individual schedulers may be constantly interacting with each other to achieve global (emergent) scheduling, exchanging information on the events that occur during the process tasks that could affect the order execution of the others. The swarm of schedulers is connected to legacy systems, namely databases and ontology services, enabling the schedulers to retrieve required data from these sources based on an example on an exchange data model (explained later). To design something similar to this, it is important to know how to define swarms to balance the interests of selfish individuals and groups of individuals. In this sense, some strategies can be implemented:

- The scheduling problem may be divided into several smaller scheduling problems according to a logical dependency; for example, a workshop scheduler can comprise several station schedulers.
- The scheduling problem may be divided in a way that different schedulers can be used to search faster alternative solutions in parallel, for example, using different scheduling algorithms or exploratory searches (optimization methods).

Both situations reinforce the importance of using swarm principles in the design of scheduling systems. The specification of individual schedulers is independent of swarm principles and can be implemented using different algorithms. In addition, individual schedulers can be implemented using multi-agent technology, composed of a set of software agents that represent interests of orders, resources, products, workers, operations and materials. In the last approach, the agents interact with each other to achieve the overall schedule through proper negotiation mechanisms, and it is possible to combine this with optimization methods through developed extensions to communicate and integrate different platforms. In this strategy, an individual scheduler may be composed of a network of entities, which are able to produce the desired scheduling where entities are related by a set of relations. The swarms of heterogeneous schedulers are supported in a transparent manner, contributing to a smooth migration from old solutions using MAS technology combined with optimization methods. For this purpose, the proper definition of communication interfaces is required as the means of data exchange between individual schedulers.

The previously presented swarm principles allow DS to obtain dynamic solutions using different types of distributed and optimized methods. This process facilitates system design, where each entity can be any application scheduler, that is, initiator, participant and optimization agent. This allows the design of a complex system using different types of agents, in particular:

- Initiator Agent—represents the "jobs" (a job represents a set of operations, processes and patients) that provide sequence and dynamic coordination to generate the process of the goals according to the scheduling and participant agents.
- Participant Agent—represents the participants of the solution (machines, health professionals, etc.). Can be responsible for generating tasks and forecasting the future generation of scheduling. They will manage their activities and behavior, considering their priorities, needs, current status and forecast of future demand. In addition, they are subject to unexpected events, interruptions or failures. For this purpose, this agent also considers

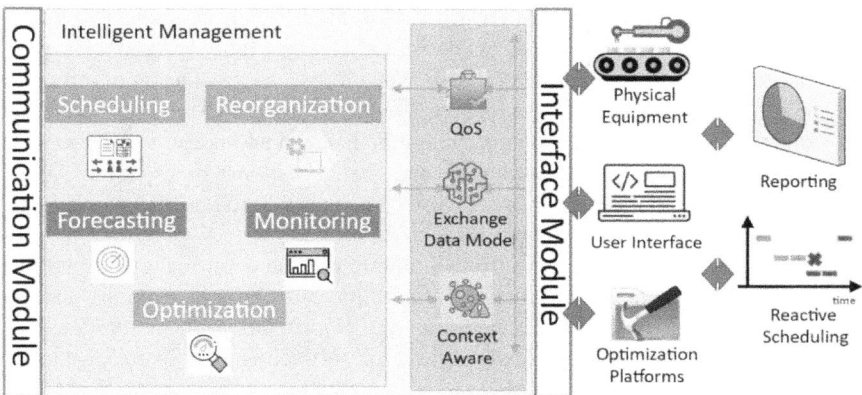

FIGURE 7.2   Architecture for intelligent management of an agent.

Source: Adapted from Alves et al., 2019c

the implementation of scheduling, system reorganization capabilities, prioritization and optimization.

- Optimization Agent—represents functional agents, that is, pieces of software used to carry out sub-tasks such as search or working strategies, local optimization or scheduling management. For example, multi-criteria, decision methods, reasoning or even AI can perform certain functions and act as a mediator.

Each agent is responsible for collecting and analyzing its data from its connected software (or a set of text files associated with the problem). This data exchange between the different agents supplies the decision-making abilities of each one, which can improve optimization procedures and scheduling activities.

In this way, the agent's conceptual architecture is illustrated in Figure 7.2, a proposal for applications in real-world environment.

Agents will be dynamically started through system administration using communication to exchange and access information, that is, with the use of a database as an engine. The multi-agent system will use cloud computing to decentralize effort tasks and later distribute intelligence based on a communication module that will manage the exchange and access of information. Smart functions relate analysis of the data from the set of tasks belonging to the DS, comprising:

- Scheduling—assesses the possibilities and performs task allocation (set of operations or processes) and management (participants involved) for each entity.
- Forecasting—forecasts the generated scheduling by using current and historical management and scheduling data, as well as disruption or failure information.

- Reorganization—represents reactive scheduling facing immense pressures and dynamic factors, such as changes in processing time, emergencies, prioritization and/or participant failures. The idea is to apply the act of modifying the offline scheduling in response to disruptions or identification of the most critical tasks for each entity and time horizon and present a predictive-reactive state in which a reschedule can start in response to a specific event, depending on the current context. It will be possible to consider, for example, new orders, failures and unexpected events.
- Monitoring—monitors the current situation and decision-making evolution to detect efficiency, reliability and flexibility, especially in dynamic procedures, such as the start of scheduling and prioritization or, in some cases, to initiate coordination and dynamics for self-scheduling in the event of disruption or failures.
- Optimization—functions to promote a decentralized system and scheduling generation that require cooperation and optimization in the use of shared resources. Optimization will be responsible for finding optimal or near-optimal solutions for local or global schedules applying specific and evolutionary optimization methods.

The design of such adaptive and self-reconfigurable scheduling systems also considers reorganization concepts to support the regulation of the dynamics of such a complex network of a swarm of schedulers. Reorganization can be seen from two perspectives: structural (changing the relationships among the individual entities) and behavioral (changing the internal behavior of individual entities). From another perspective, structural self-organization appears when:

- Dynamic reconfiguration of the scheduling problem occurs, such as by adding/removing tasks, resources or workers or by changing the dependencies between the individual schedulers inside the swarm scope.
- A dynamic reorganization of individual entities involved in exploratory searches for planning or scheduling solutions occurs, aiming to achieve better emergent solutions.

The intelligent management will address three small assistance and support modules, which in turn will communicate with the interface module in a bi-directional approach. Relative to the exchange data model, a common process model is required, such as the entering of input data and the obtained solution (output) to the problems caused by the architecture and connections, which includes different problems, constraints, restrictions and methods. On the other hand, the interface module can be connected to physical devices and optimization platforms (e.g., dedicated platforms/algorithms embedded), access a dashboard dedicated to monitoring and interact with a specific optimization platform or communication service.

The optimization platform module is responsible for agent behavior and system performance and is provided by a dedicated platform embedded with algorithms and optimization methods, such as MATLAB, ILOG Cplex or Python. Thus, it will

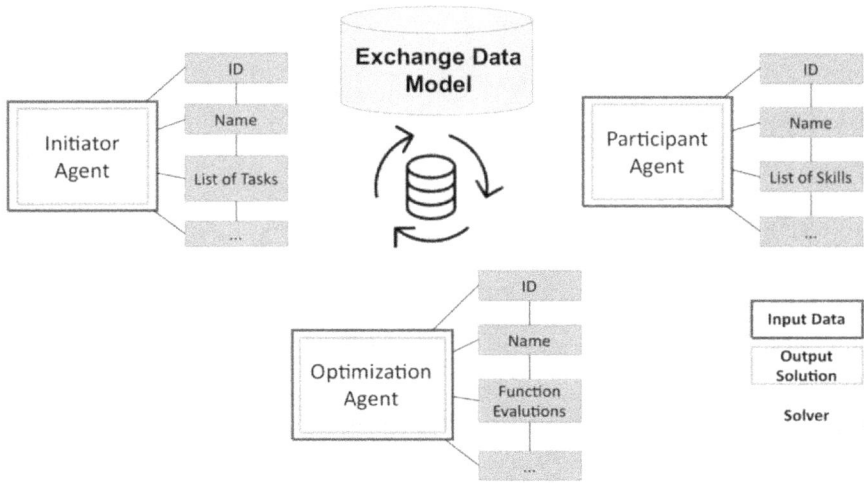

FIGURE 7.3   General exchange data model.

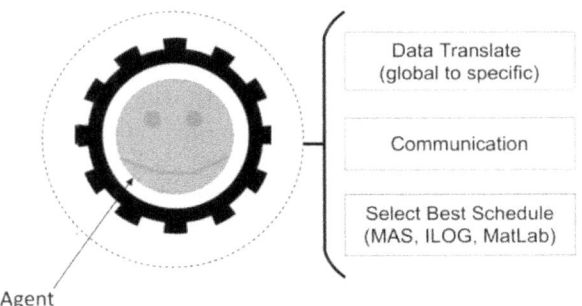

FIGURE 7.4   Capabilities of each agent.

be possible to obtain cooperation and communication between platforms, especially dedicated to specific actions, enhancing their resources and libraries.

Resulting from the combination of MAS and optimization methods, the data format, its semantics and/or its variables needs to be standardized. Figure 7.3 illustrates a general model of data exchange for any application of scheduling in an attempt to standardize the data and the consequent exchange of information.

Each agent will have at its disposal a data model capable of being applied in different domains. In turn, the conceptual structure will be supported depending on the type of agent and will be developed with heterogeneous perspectives in the construction data sources for dynamic environments (Figure 7.4).

In this sense, the modules inside each agent will be able not only to adjust their own parameters but also to adjust according to the result of interaction with other agents. This means that each agent has the ability to communicate, interact and

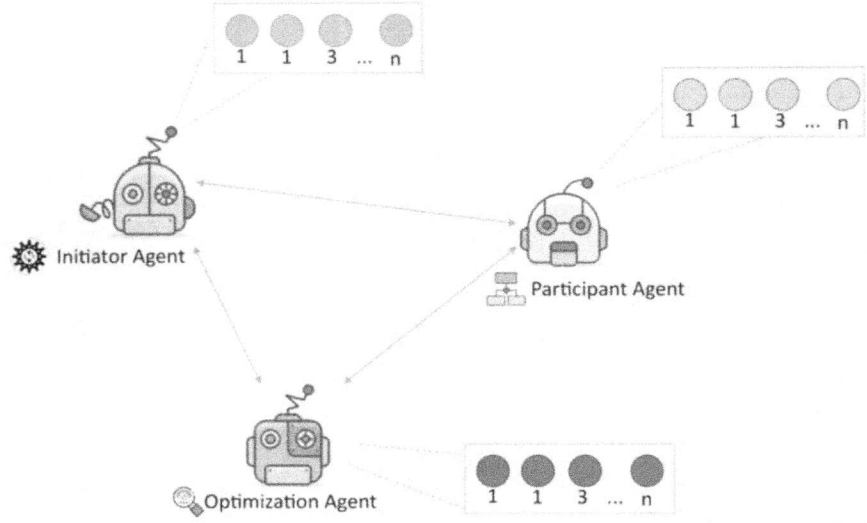

FIGURE 7.5   Interactions among agents.

cooperate with other agents, using in this case the communication module, following proper coordination patterns that specify how and which interaction pattern should be used based upon a specific context. The communication protocol will be mainly from agent to agent, preferentially using the FIPA-ACL protocol. In this way, each entity can select specific agents and interact with them. Figure 7.5 represents generic interactions between the aforementioned agents. These interactions will be modelled using Agent Unified Modelling Language (AUML) (Bauer et al., 2001) that extends the UML to design agents through an intuitive graphical representation of the architecture and processes.

The intra-agent interface module is responsible for providing mechanisms that make access to the optimization platforms transparent, for example, MATLAB or other software, supporting the collection of data and the execution of commands in the distributed system, such as the scheduling of set tasks. In general, the proposed disruptive architecture will empower decision-making for DS in real applications with dynamic environments, providing competitive and intelligent technological skills in reporting cases and strategic and automated planning for reactive scheduling problems.

7.4   CASE STUDIES

The proposed approach can be applied to different domains aiming to provide environments to validate the proposal's benefits. Two application domains are considered for this study: one in the manufacturing area and the other in the home health care (HHC) area. At this stage, it is not clear which domain can benefit the most from this work. Therefore, the following methodology seeks to validate the conceptual MAS approach, as illustrated in Figure 7.6.

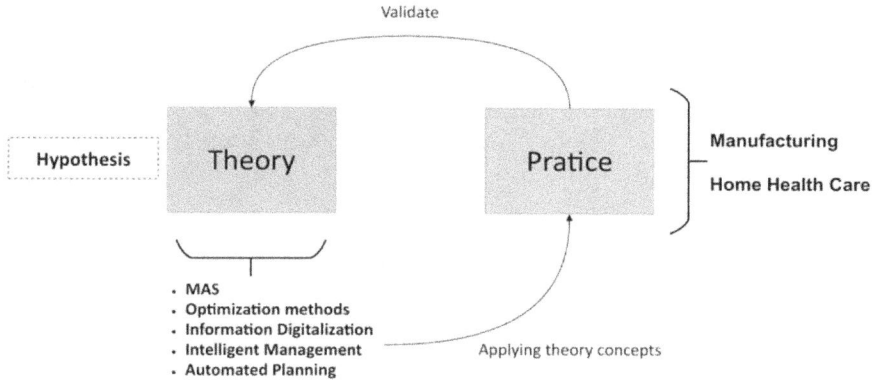

FIGURE 7.6   Validation methodology.

7.4.1   MANUFACTURING

The case study based on the manufacturing domain, namely the flexible manufacturing system production cell AIP PRIMECA (Trentesaux et al., 2013), has a set of several robotic workstations. A set of operations (i.e., jobs) are performed on the workstations, generating different products that, in turn, have a production workflow according to their sequence of operations until the final product for the consumer is obtained. In turn, these products can be performed on different workstations depending on their skills; however, each station has different performance. The goal concerns the reduction of the re-planning time considering an increase in the quality of the services provided under very dynamic constraints, for example, the occurrence of unexpected events or the need for reconfiguration. Obviously, these problems fit into several fields of research, such as critical path analysis, supply chain management and resource allocation, among others. Note that the idea is to avoid using centralized approaches but instead to use innovative optimization methods provided by combining with MAS. For this case study, it is useful to understand what (and how) our proposal can offer for this particular case, enabling a resolution of problems to:

- Enable and prioritize task effort distribution.
- Reduce the possibilities of interrupted processes.
- Achieve faster and possibly more accurate reaction to condition changes.

This case study intends to highlight the advantages of promoting a reactive and flexible architecture for distributed scheduling, seeking, whenever possible, to balance the workload of those involved through a conceptual and simulated model that evaluates, in runtime, several replacement options and proposals for new scheduling in case of necessity. As new proposals emerge, the best ones will be selected and will support the agent's organization.

Some approaches and applications have already been developed in the manufacturing domain involving AIP PRIMECA and similar cases. The dataset was generated using NetLogo, a multi-agent programmable modelling environment, powered by agent-based simulations and proper validation for the case study (Alves et al., 2018b; Alves et al., 2019d). The experimental results achieved dynamic responsiveness for distributed scheduling with fast responses to changes in the environment and supporting decision-making in an interactive and user-friendly way.

7.4.2 HOME HEALTH CARE

A home health care support service allows the provision of care (medical, nursing or social) at the home of the user (Fikar and Hirsch, 2017). HHC has gained importance in recent years, particularly in Portugal, as it allows users to travel less and have more comfort in their treatment (of particular importance given the aging of the population). The management of HHC has a strong component related to the adequate use of resources (doctors, nurses, social workers, vehicles) in the fulfilment of user treatment (typically characterized in spatial and temporal terms and also by the specific requirements of the caregivers). The focus is to enable decision-making at an operational level, where as a rule it is up to the health professional who will provide the HHC to ensure the schedule and transport conditions for it, which currently does not have optimized circuits of visits or any computational support. In the current HHC management system (usually manual), it is often necessary to make decisions based on priority principles of care, leaving aside some care due to lack of time or excessive workload for healthcare professionals. In addition, the frequency of some HHC does not respect the regularity imposed due to flaws in the scheduling channels and the lack of time that professionals have for proper planning. Resource management, such as vehicles and routes, is often scarce, and hospital or health unit management is regulated in terms of access and proper functioning, even in the case of unexpected events that require real-time action (Cissé et al., 2017). Typically, traffic jams, road accidents, weather conditions, new orders and emergencies should be considered when designing a conceptual architecture capable of supporting a complex optimization system, with several constraints and restrictions (Rasmussen et al., 2012).

The main idea is to consider the problem of assigning tasks outside a health unit, belonging to the National Health System, or applied in HHC belonging to a social solidarity institution. The approach will be in a first instance, and it will be to modernize the current HHC system, digitizing the information and enabling an information system with intelligent and optimized decision-making. In a second phase, it will be a priority to find the optimal scheduling for visits and home care circuits, especially considering complex professional assignments, routes and workload balancing for a given day, week or specific time horizon. In addition, the maintenance and monitoring of correct continuity of care will be predominant. Therefore, it will be important to apply the proposed methodology in an attempt to decompose the problem into sub-problems (e.g., division in health units, daily and/or monthly scheduling, traveling issues, workload balancing) and provide distributed schedules with local optimization to be inserted and included in the overall scheduling or allocation.

Some approaches and applications in the HHC domain have already been developed involving a health unit in a real implementation environment with a specific real dataset (Alves et al., 2019a; Alves et al., 2019b; Alves et al., 2017). However, in recent times, studies point to the need to introduce intelligent systems with fast reaction in dynamic environments, mainly due to the lack of flexibility of most classical optimization techniques. For that reason, MAS has already been introduced to take advantage of cooperation and intelligent mechanisms to decentralize the system, which normally operates in static mode. In a first approach, a simulated methodology in NetLogo provided excellent capacity to generate good autonomous and coordinated solutions, contributing to a better decision-making and usefulness (Alves et al., 2018a). In the future, the idea is to contribute to an HHC system based on a digital and distributed ecosystem (platform/app) using different ICT-based solutions, sources, tools and techniques.

7.4.3 EVALUATION AND PERFORMANCE MEASUREMENTS

Regardless of the case study, the proposed approach should be evaluated and validated in relation to some main aspects and features. The comparison of different existing approaches and applications is a very complex task since they present specific functionalities, purposes and other aspects that can mean it requires too much time and effort to obtain comparison measurements. Therefore, regarding the comparison between the proposed approach and other existing approaches and applications, a more qualitative evaluation will be performed. In addition, quantitative assessments will be carried out through the analysis and measurement of behaviors of the case study system prototype. The evaluation process will consist of the assessment and quantification of the results derived from the system prototype execution in the case study scenarios, considering the response time, output accuracy and computational resource utilization or effort the main metrics. In this sense, since the proposed approach intends to cover multi-agent system design for DS, the first aspect that should be evaluated is the local distributed vs. centralized monitoring, using as metrics the response time and monitoring output accuracy. In order to simplify the experiments, computational resource requirements (such as processing capacity and memory) can be fixed during the quantitative analysis and evaluated in a qualitative manner considering the input information, variables and constraints.

The configuration of the local and distributed experiment will consist of several agents using communication mechanisms and optimization methods for collaboration and individual monitoring of different system components. The results of such a configuration, in terms of response time and accuracy, should be compared with the results of a centralized configuration, where the data of all components should be sent to a central repository and then analyzed to obtain a global schedule. In this experiment, a distributed diagnosis should be achieved by the agents' interaction, while the centralized diagnosis will be performed by only one agent. Thus, the distributed approach is expected to require more network resources, while the centralized approach should require more memory. The accuracy of the diagnosis presented by both approaches should be equivalent, and the time should not differ too much. In this sense, the distributed approach may be qualitatively better based on its flexibility and adaptability in autonomous scheduling.

In distributed scheduling, the society of agents will be able to receive requests and respond according to their specifications, seeking, whenever possible, to find and select the agents with the best proposals to be allocated the scheduling. The data should be compared with the initial schedule as one of the hypotheses that measures expectations about a specific feature. The main goal, in general, is to try to automatically compose the best planning of tasks, according to the requirements and objectives in question, making it possible to make statements by forwarding and receiving messages. Obviously, in the case of experiment disruptions or emergencies and the need for self-scheduling, the collected data should be analysed to make possible a decision, with minimal human intervention, on the feedback using qualitative indicators (e.g., flexibility, robustness, agility, etc.) and measure the following quantitative key performance indicators (KPIs):

- Scheduling composition (maximum time, distance or other objectives).
- Quality of composition (average quality of the scheduling composition played inside the swarm agents).
- Proposal negotiation and agent utility (how well an agent can offer a schedule and if the best offer is chosen).
- Average communication load (exchanged information in the system).
- Average number of agents (average number of agents required to compose a certain schedule or task request).
- Average forwarding messages (average cost of dispatching requests to other agents).
- Resilience of the network (ability to remove agents from the network without incurring a loss of performance).
- Average message size (exchanged message between the agents).

In Table 7.1, the evaluation intended to be performed is summarized.

The analysis of the solutions, composition, runtime, and statistical parameters will allow to obtain information about all the system, its modules, and procedures, to support decision-making. Thus, by using this information and measuring its impact on the overall system performance and how this has contributed to improving the dynamic adaptation, reconfiguration of processes, efficiency, effectiveness, trade-offs, and robustness of conceptual MAS for DS, will be a great approach to take into account.

7.5 CONCLUSIONS AND FUTURE WORK

This work points out and discusses a MAS infrastructure for distributed scheduling and resource allocation that is able to adapt and evolve using DAI (i.e., a mechanism of relationships among agents) and fast decisions in dynamic and reactive environments. All the conceptual and technical aspects allow for a robust, intelligent and powerful control approach characterized by the MAS's capabilities to provide adaptation and fast responses to address scheduling and allocation challenges for multidisciplinary domains. Thus, the system is able to adapt and evolve using structural self-scheduling (i.e., mechanisms of relationships among agents), optimized solutions and fast decisions in dynamic and reactive environments. Two case studies focusing

TABLE 7.1

Summary of the Proposed Approach Evaluation and Metrics

Description	Metric	Expected Results
Compare, in a qualitative manner, the proposed approach with other existing approaches and applications	Relevant functionalities and features required and desired to attend to the current industrial and health issues and challenges	Cover most of the relevant aspects and features
Evaluate the monitoring and diagnosis tasks considering the proposed distributed swarm vs. a centralized approach	Response time, response accuracy, resource constraints	Obtain better response time, with acceptable accuracy, considering resource variables, constraints and requirements
Compare, in a quantitative manner, the MASs	Scalability and flexibility (measure the performance of all previous KPIs)	Cooperating and evolve in order to reduce or increase several points (e.g., decrease schedule composition costs and improve utility)
Evaluate solutions, runtime and statistical parameters in the optimization methods	Parameters, execution time, statistical analysis, quality, computational effort of execution and robustness	Adjust the parameters, solving problems in a reasonable time; guarantee the quality of the "optimal" solution; and apply different problems and instances

on distributed and dynamic environments were considered. The focus for the future will be on the experimental approach in an attempt to potentiate an architecture for distributed scheduling resolution. Due to the dynamics and resources of the system concerned, performance measures and metrics will be crucial for the evaluation of the prototype.

In order to properly address scheduling challenges, it is expected a robust, intelligent, and powerful control approach characterized by the combination of MAS and optimization methods with swarm capabilities. In addition, intelligent and resilient mechanisms, namely forecasting and self-scheduling, will offer the possibility of developing and testing algorithms and strategies for context-aware supervision and monitoring that can contribute to more effective system adaptation. Distributed scheduling can be specified by a decentralized approach with intelligent services and procedures to support a lack of computational availability and reactive mechanisms. The conceptual MAS design for the DS system showed that it is a promising approach, but still, the validation of the current implementation is mandatory as future work to extrapolate the results and analyze the principles used in terms of evaluation and performance measures, namely the KPI. In this way, the DS system can improve existing results with responsiveness to dynamic data, decentralization, and optimization in domains with emerging needs, with great prospects for applicability, flexibility and scalability.

Summing up, an approach was proposed to explore the digitization of information and cloud support for user-friendly use that will allow exponential applicability and aid decision-making in dynamic environments. The system realized a MAS approach with several intelligent capabilities and optimization algorithms will allow its portability and authenticity in a scalable and flexible infrastructure for monitored management, even in real-time scheduling environments.

ACKNOWLEDGMENTS

This work has been supported by FCT—Fundação para a Ciência e a Tecnologia within the R&D Units Projects Scope: UIDB/00319/2020 and UIDB/05757/2020. Filipe Alves is supported by FCT Doctorate Grant Reference SFRH/BD/143745/2019.

NOTES

1 http://fipa.org/
2 http://jade.tilab.com/
3 https://ccl.northwestern.edu/netlogo/

REFERENCES

Alves, F., Alvelos, F. P., Rocha, A. M. A., Pereira, A. I., and Leitão, P., (2019a). Periodic vehicle routing problem in a health unit. *Proceedings of the 8th International Conference on Operations Research and Enterprise Systems—ICORES,* pp. 384–389.

Alves, F., Costa, L., Rocha, A. M. A., Pereira, A. I., and Leitão, P., (2019b). A multi-objective approach to the optimization of home care visits scheduling. *Proceedings of the 8th International Conference on Operations Research and Enterprise Systems—ICORES,* pp. 435–442.

Alves, F., Pereira, A. I., Barbosa, J., and Leitão, P., (2018a). Scheduling of home health care services based on multi-agent systems. *International Conference on Practical Applications of Agents and Multi-agent Systems,* pp. 12–23.

Alves, F., Pereira, A. I., Fernandes, F. P., Fernandes, A., Leitão, P., and Martins, A., (2017). Optimal schedule of home care visits for a health care center. *International Conference on Computational Science and its Applications,* pp. 135–147.

Alves, F., Rocha, A. M. A., Pereira, A. I., and Leitão, P., (2019c). Distributed scheduling based on multi-agent systems and optimization methods. *International Conference on Practical Applications of Agents and Multi-agent Systems,* pp. 313–317.

Alves, F., Varela, M. L. R., Rocha, A. M. A., Pereira, A. I., Barbosa, J., and Leitão, P., (2018b). Hybrid system for simultaneous job shop scheduling and layout optimization based on multi-agents and genetic algorithm. *International Conference on Hybrid Intelligent Systems,* pp. 387–397.

Alves, F., Varela, M. L. R., Rocha, A. M. A., Pereira, A. I., and Leitão, P., (2019d). A human centred hybrid MAS and meta-heuristics-based system for simultaneously supporting scheduling and plant layout adjustment. *FME Transactions,* pp. 699–710.

Baker, K. R., and Trietsch, D., (2013). *Principles of sequencing and scheduling.* John Wiley & Sons.

Bartz-Beielstein, T., Parsopoulos, K. E., and Vrahatis, M. N., (2004). Analysis of particle swarm optimization using computational statistics. *Proceedings of the International Conference of Numerical Analysis and Applied Mathematics* (ICNAAM 2004), pp. 34–37.

Bauer, B., Müller, J. P., and Odell, J., (2001). Agent UML: A formalism for specifying multi-agent software systems. *International Journal of Software Engineering and Knowledge Engineering,* 11(03), pp. 207–230.

Behnamian, J., and Ghomi, S. F., (2016). A survey of multi-factory scheduling. *Journal of Intelligent Manufacturing*, 27(1), pp. 231–249.

Bonabeau, E., Marco, D. D. R. D. F., Dorigo, M., Théraulaz, G., and Theraulaz, G., (1999). *Swarm intelligence: From natural to artificial systems* (No. 1). Oxford University Press.

Bordini, R. H., Braubach, L., Dastani, M., Seghrouchni, A. E. F., Gomez-Sanz, J. J., Leite, J., and Ricci, A., (2006). A survey of programming languages and platforms for multi-agent systems. *Informatica*, 30(1).

Cardoen, B., Demeulemeester, E., and Beliën, J., (2010). Operating room planning and scheduling: A literature review. *European Journal of Operational Research*, 201(3), pp. 921–932.

Chan, H. K., and Chung, S. H., (2013). *Optimisation approaches for distributed scheduling problems*. Taylor & Francis.

Cissé, M., Yalçındağ, S., Kergosien, Y., Şahin, E., Lenté, C., and Matta, A., (2017). OR problems related to home health care: A review of relevant routing and scheduling problems. *Operations Research for Health Care*, 13, pp. 1–22.

Ferber, J., (1999). *Multi-agent systems: An introduction to distributed artificial intelligence* (1st ed.). Addison-Wesley Longman Publishing Co., Inc.

Fikar, C., and Hirsch, P., (2017). Home health care routing and scheduling: A review. *Computers & Operations Research*, 77, pp. 86–95.

Finin, T., Fritzson, R., McKay, D., and McEntire, R., (1994). KQML as an agent communication language. *Proceedings of the Third International Conference on Information and Knowledge Management*, pp. 456–463.

Jabeur, N., Al-Belushi, T., Mbarki, M., and Gharrad, H., (2017). Toward leveraging smart logistics collaboration with a multi-agent system based solution. *Procedia Computer Science*, 109, pp. 672–679.

Kravari, K., and Bassiliades, N., (2015). A survey of agent platforms. *Journal of Artificial Societies and Social Simulation*, 18(1), pp. 11.

Labrou, Y., Finin, T., and Peng, Y., (1999). Agent communication languages: The current landscape. *IEEE Intelligent Systems and Their Applications*, 14(2), pp. 45–52.

Leitão, P., (2009). Agent-based distributed manufacturing control: A state-of-the-art survey. *Engineering Applications of Artificial Intelligence*, 22(7), pp. 979–991.

Leitão, P., Mařík, V., and Vrba, P., (2012). Past, present, and future of industrial agent applications. *IEEE Transactions on Industrial Informatics*, 9(4), pp. 2360–2372.

Leung, J. Y., (2004). *Handbook of scheduling: Algorithms, models, and performance analysis*. CRC Press.

Lezama, F., Palominos, J., Rodríguez-González, A. Y., Farinelli, A., and de Cote, E. M., (2019). Agent-based microgrid scheduling: An ICT perspective. *Mobile Networks and Applications*, 24(5), pp. 1682–1698.

Marcon, E., Chaabane, S., Sallez, Y., Bonte, T., and Trentesaux, D., (2017). A multi-agent system based on reactive decision rules for solving the caregiver routing problem in home health care. *Simulation Modelling Practice and Theory*, 74, pp. 134–151.

Michael, L. P., (2018). *Scheduling: Theory, algorithms, and systems*. Springer.

Miller, P., (2007). The genius of swarms. *National Geographic*, 212(1), pp. 126–147.

Monostori, L., Váncza, J., and Kumara, S. R., (2006). Agent-based systems for manufacturing. *CIRP Annals*, 55(2), pp. 697–720.

Nealon, J. L., (2003). *Applications of software agent technology in the health care domain*. Springer Science & Business Media.

Pinedo, M., Zacharias, C., and Zhu, N., (2015). Scheduling in the service industries: An overview. *Journal of Systems Science and Systems Engineering*, 24(1), pp. 1–48.

Rasmussen, M. S., Justesen, T., Dohn, A., and Larsen, J., (2012). The home care crew scheduling problem: Preference-based visit clustering and temporal dependencies. *European Journal of Operational Research*, 219(3), pp. 598–610.

Russell, S., and Norvig, P., (2002). *Artificial intelligence: A modern approach.* Prentice Hall.

Singh, M. P., (1998). Agent communication languages: Rethinking the principles. *Computer*, 31(12), pp. 40–47.

Smith, R. G., (1980). The contract net protocol: High-level communication and control in a distributed problem solver. *IEEE Computer Architecture Letters*, 29(12), pp. 1104–1113.

Trentesaux, D., Pach, C., Bekrar, A., Sallez, Y., Berger, T., Bonte, T., Leitão, P., and Barbosa, J., (2013). Benchmarking flexible job-shop scheduling and control systems. *Control Engineering Practice*, 21(9), pp. 1204–1225.

Van den Bergh, J., Beliën, J., De Bruecker, P., Demeulemeester, E., and De Boeck, L., (2013). Personnel scheduling: A literature review. *European Journal of Operational Research*, 226(3), pp. 367–385.

Wang, S., Wan, J., Zhang, D., Li, D., and Zhang, C., (2016). Towards smart factory for Industry 4.0: A self-organized multi-agent system with big data based feedback and coordination. *Computer Networks*, 101, pp. 158–168.

Weiss, G., (1999). *Multiagent systems: A modern approach to distributed artificial intelligence.* MIT Press.

Wooldridge, M., (1997). Agent-based software engineering. *IEE Proceedings-software*, 144(1), pp. 26–37.

Wooldridge, M., (2009). *An introduction to multiagent systems.* John Wiley & Sons.

Wooldridge, M. J., and Jennings, N. R., (1995). Intelligent agents: Theory and practice. *The Knowledge Engineering Review*, 10(2), pp. 115–152.

8 Environmental Impact Assessment during Additive Manufacturing Production
Opportunities for Sustainability and Industry 4.0

Alok Yadav, Anbesh Jamwal, Rajeev Agrawal,
Vijaya Kumar Manupati and Jose Machado

CONTENTS

8.1 INTRODUCTION

Industry 4.0, sometimes known as the fourth industrial revolution, is one of the most promising production revolutions in history [1]. Additive manufacturing (AM) is a critical enabler for the fourth industrial revolution, and it has the potential to have a

DOI: 10.1201/9781003123866-8

long-term impact on the production environment [2]. AM represents a paradigm shift in manufacturing industries regarding how we make parts and products customizable for customers in the minimum amount of time [3]. This technique has the potential to increase manufacturing sustainability by lowering material use, developing new shapes and configurable designs, and cutting production times, according to reports and published literature. This is altering industries so that new business models can be adopted, while completely integrating AM into an industrial environment will necessitate a certain level of design expertise. It has the potential to improve the manufacturing industry by lowering production, distribution, and inventory costs, as well as facilitating the development and industrialization of new products [5]. Organizational strategy is guided by the market's progress and the rapid adoption of emerging technology. This chapter seeks to contribute to the computation of the impacts of AM methods on sustainable business models by using the lenses of Industry 4.0 and its technological principles.

Industry 4.0 has reformed the world in various ways, one of them being the generation of opportunities for sustainable manufacturing and development [6]. The introduction of these technologies, for example, has advanced manufacturing technologies and provides us with lower material wastage and the ability to fulfill make-to-order by AM.

AM is a technology or collection of ideas that make a product by adding material layer upon layer in an exact geometrical shape [7]. AM can be considered the most disruptive innovation in the present manufacturing world, as it helps to change value chains from the design cycle to the end of life and gives benefits over traditional manufacturing regarding both flexibility and waste minimization. AM is a very significant process, as markets are quickly changing and customers are asking for more different and customized products. Impact assessment in AM is still a limited research area [7, 8]. Now industries are adopting AM practices to promote Industry 4.0 practices [9]. There is a need to study the environmental impact generated from AM processes. To fill this gap, we have conducted an analysis on AM products. The research concerns of the present chapter are:

> RQ1: To assess the environmental impact of pattern making by both conventional and AM processes.
> RQ2: Selection of the best environmentally sustainable process based on results.

To achieve these objectives, a real-time case study is performed in small and medium enterprises in India. The data is collected by questionnaire survey and semi-structured interviews. The impact assessment is done by GABI software on the Windows 10 operating system. In the next sections, we discuss the literature review, methodology used, results and discussion, and opportunities for AM and sustainability in Industry 4.0.

8.2 LITERATURE REVIEW

There are works in the literature that compare AM technology to the methods that are commonly used for the manufacturing of complex parts or patterns, referred to as the conventional manufacturing method. One of the most exciting manufacturing developments in the Fourth Industrial Revolution is AM. Because of the advancement of technology and the constraints imposed by COVID-19, in numerous aspects, we can certainly declare that we have reached a tipping point when AM is set to become a viable alternative to traditional manufacturing procedures. AM isn't just for making new models; it can also be used to make things with better quality, less waste, and a production pace that can be adapted to the project's demands. AM's rising popularity originates from its capacity to manufacture in a flexible manner with optimized logistics, hence bringing up new business prospects. The automotive, aerospace, and defense industries were among the first to employ AM in prototyping, with benefits such as product customization, development of complex tools and components, reduction of time and cost to market, and creation of new designs and lighter products. According to a study done in Europe, demand for AM will reach 11 billion euros in the next decade, fueled by sales of prototypes, goods, and component production, with a target of 130 billion euros in the next decade. AM, when paired with artificial intelligence and the Internet of Things, is revolutionizing the healthcare business, allowing for huge breakthroughs in medical technology and surgical procedures like orthopedics, dental, pharma, cardiology, and tissue engineering.

Industries that utilize or plan to use this new technology are interested in reaping not just economic gains but also strategic advantages in the industry, as well as social and environmental benefits. Changes in the value chain's structure are being driven by consumer demand for individualized and sustainable production, as well as environmental regulations. One of the major challenges in implementing AM technology, as with any emerging technology, is a lack of understanding of the effects of technology adoption on supply chain reconfigurations and the implementation of business models that are increasingly based on sustainability goals.

The modern industrial revolution is related to the digital world, and AM is a part of it. It denotes a completely new method of manufacturing that involves layering materials before the desired result is achieved [10, 11]. Long term, this new manufacturing model will allow us to eliminate numerous manufacturing tools and modify production processes to consumer needs with far better flexibility [12, 13]. In various aspects, recent technological advancements indicate that AM is on the verge of becoming a viable alternative to traditional manufacturing procedures [14, 15]. Materials used, direct or indirect process technology, and the state of the raw material used have all been used to classify AM processes in the literature. Previous research has revealed major advancements in AM technology in terms of applications, economic effects, raw material use, and design. Filament and powder are, to our knowledge, the most commonly used raw materials for cost characterization, business model ideas, and sustainability processes. Furthermore, as opposed to conventional manufacturing methods, filament and powder are the most popular AM technologies used for Industry 4.0 integration [16].

8.3 METHODOLOGY

Sand casting is a conventional manufacturing process that has been used for the last 5000 years [17]. During this time, modifications have been made to this process according to advancements in technology. To calculate the emissions produced by the sand-casting process, life cycle assessment (LCA) is required [18]. LCA is a systematic process that is performed step by step [19]. We need to know the input and output of each process with the correct quantities. Product details, system boundaries, and functions should be fixed before the study. Almost two-thirds of metal casting products are produced by sand casting, including parts of molten metal, usually stainless-steel alloys, brass, Al, and carbon steel [20, 21].

8.3.1 Needed Improvements in Sand Casting

Emissions from sand casting can be reduced by either by process modification or additive modification, but it is necessary to know if the new process is useful.

Currently, 3D printing is used in sand casting and replaces traditional processes, but it is necessary to know whether 3D printing is beneficial from various perspectives; specifically, for the case of sand casting, the environmental impact factors, solid or liquid waste considerations, and energy requirements all need to be studied so that there is transparency of emission levels. All these factors can be found by LCA; its result provides a clear picture of emission levels to help industry select better processes that optimize the quality of finished goods with a lower level of emissions.

AM currently plays a big role in manufacturing industries [22]. AM includes a wide range of applications. 3D printing is one of them; it is used for rapid-investment casting for pattern making [23]. The AM process is a three-step process. First is CAD file formation, second is converting the standard triangle language file to read to the 3D printer, and last is the final output. The emission level varies in 3D printing with different materials. It may be higher or lower than sand casting [24]. Detailed research work is required to compare 3D printing emission levels.

8.3.2 3D Printing in the Sand-Casting Process

AM plays an important role in the present scenario. AM includes a wide variety of applications, one of which is 3D printing for rapid sand casting. There are different types of 3D printing technologies, and each includes specific strengths and weaknesses. Today, 3D printing is used in sand casting for pattern making [25, 26]. Different types of material used in 3D printing have different chemical or physical properties.

Overall, 3D printing is used in many industries. In the future, 3D printing may replace many traditional manufacturing processes. So, it is also necessary to study the environmental emissions created by each 3D printing process. Emission levels by 3D printing processes vary according to the material used in the process. Emissions may be higher or lower than those from the sand-casting process. A complete, detailed research work would help to find the strengths and weaknesses of the 3D printing process. Also, this will help industry gain knowledge regarding 3D printing emissions.

8.3.3 Product Details

The aluminum alloy used in this research work is $Al\text{-}Si_5\text{-}Cu_3$. This alloy has satisfactory casting properties and a wide range of applications in different manufacturing processes.

In this research work, LCA performs as a decision tool to choose the best available technique for sand casting.

8.3.4 PATTERN MAKING BY 3D PRINTING PROCESS

The final product is created using a 3D printer in this technique. The polylactic acid (PLA) is used as the raw material in this procedure to create the pattern. PLA is a thermoplastic aliphatic polyester that is made from renewable resources. The raw material for the 3D printer is 0.10 kg of PLA, and the end result weighs 0.09 kg.

8.4 RESULTS AND DISCUSSION

The data for this research work is collected from a company (XYZ) situated in Tamil Nadu (a state of India).

1. Questionnaires were sent via email to fill out by company.
2. Interviews with managers and employees were conducted.

8.4.1 INVENTORY FLOW AND ALTERNATIVES TO THE SAND-CASTING PROCESS

The amount of energy input, raw material consumption, and alternative energy input for the sand casting process is given in Table 8.1.

The quantity of emissions using the energy inventory from different scenarios with the help of GABI software is shown in Table 8.2.

8.4.2 SAND-CASTING MODEL WITH PATTERN MAKING BY 3D PRINTING

This model considers the application of 3D printing in the sand-casting process. Here 3D printing is used to produce the pattern for the sand-casting process. The material used is PLA. The LCA studies performed to compare the environmental impact of

---

TABLE 8.1
Inventory Flow for Sand Casting

Scenarios	Raw Material (in kg)	Consumption of Energy (in kWh)
S1	33.01	5.5
S2	33.01	5.5
S3	33.01	5.5
S4	33.01	5.5
S5	33.01	5.5
S6	33.01	5.5
S7	33.01	5.5
S8	33.01	5.5

TABLE 8.2

Details on Emissions for Each Scenario

S	GWP	AP	EP	ODP	FAETP	HTP	POCP
S1	14	0.081	4.8	4.2	0.073	27	4.85
S2	8.01	0.028	2	8.6	0.061	26	2.1
S3	8.09	0.029	2.01	11	0.064	26.2	2.15
S4	8.18	0.05	5.5	3.5	0.080	26.4	5.3
S5	7.9	0.021	1.98	3.4	0.053	25.9	2.1
S6	13	0.082	4.8	4.07	0.065	27	4.8
S7	13.1	0.075	4.85	4.2	0.066	27	4.85
S8	9.5		3.87	3.8	0.051	14	4.1

S: Scenario; **S1:** Base scenario; **S2:** Wind energy; **S3:** Solar energy; **S4:** Energy by biomass; **S5:** Energy by hydropower; **S6:** Reduction in design consumption; **S7:** Waste sand re-use; **S8:** Recycling of scraps; **GWP:** Global warming potential; **AP:** Acidification potential; **EP:** Eutrophication potential; **ODP:** Ozone depletion potential; **FAETP:** Fresh water ecotoxicity potential; **HTP:** Human toxicity potential; **POCP:** Photochemical ozone creation potential

3D printing and machining to produce a pattern for the sand-casting process show that 3D printing is good in terms of lower environmental emissions. The major factor for this reduction in emissions is the use of PLA as a raw material in 3D printing as compared to aluminum, because between the production processes of aluminum ingots and PLA, it is found that the production process of aluminum requires more energy and produces more emissions compared to PLA production. PLA, on the other hand, has a significant negative impact on human health and ecological quality. Indeed, intensive agriculture with the use of many chemicals (pesticides, herbicides, and fertilizers) that are detrimental to persons and the environment is required for the production of raw material. The production of new Al results in around 1% of global annual greenhouse gas (GHG) emissions. Mining, refining, smelting, and casting primary Al releases about 0.4 billion tons of carbon dioxide equivalent (CO_2) emissions per year. Overall, PLA produces much fewer emissions compared to the aluminum production process.

Aluminum and PLA are recyclable materials, and aluminum is an infinitely recyclable material, and from this point of view, both are good. PLA is biodegradable under industrial composting conditions, starting with a chemical hydrolysis process, followed by microbial digestion to ultimately degrade the PLA. Also, incineration can be used for PLA, leaving no residue, and it produces 19.5 megajoules per kilogram. Figure 8.1 shows the molding process performed using GABI software.

Table 8.3 shows the input data (raw material and energy) and output in the form of emissions and solid waste by using a 3D printer.

Table 8.4 shows the quantity of emissions produced as found by GABI software by using the inventory data shown in Table 8.3.

Figure 8.2 shows the results obtained from GABI software during pattern making using a 3D printer.

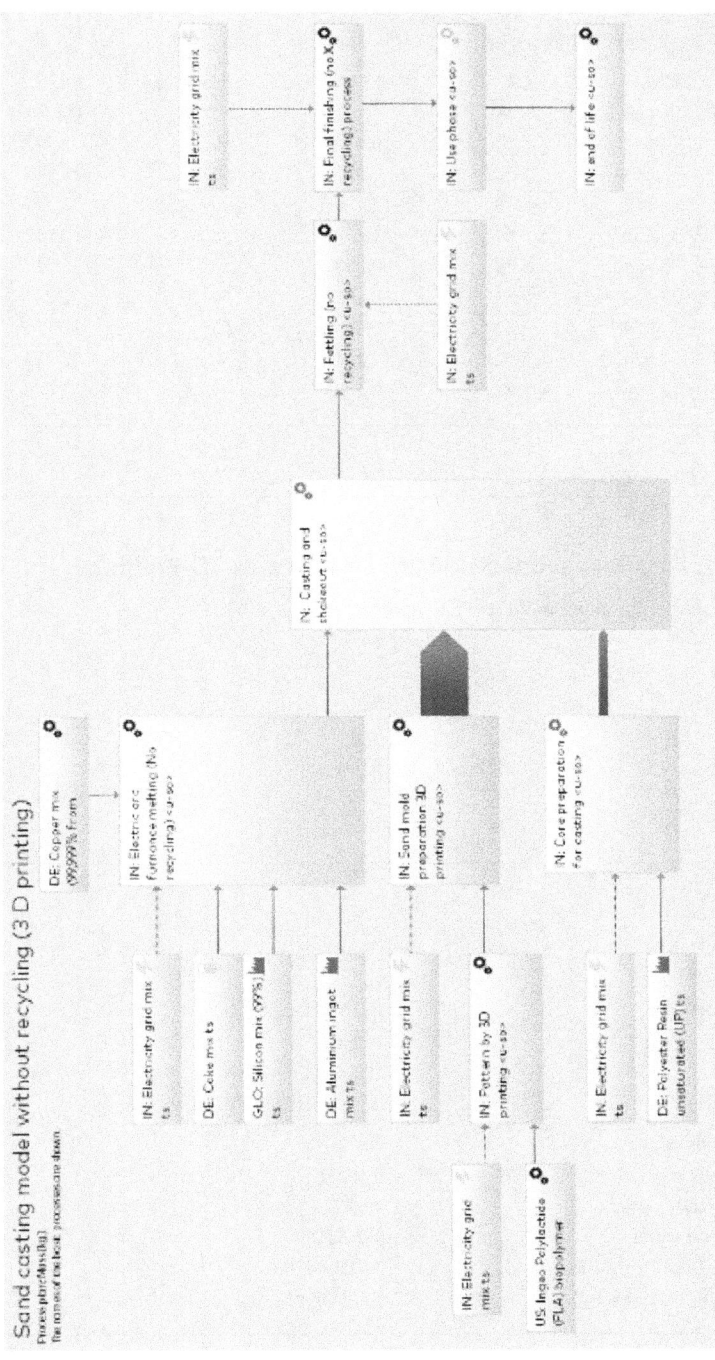

FIGURE 8.1 GABI model (pattern making by 3D printer).

TABLE 8.3

Inventory Data for 3D Printer

S	URM (Kg/FU)	UE (Kwh/FU)	ETA (%)	ETW (%)	SW (Kg/FU)
S1	33.654	5.52	0.234	50	32.717
S2	33.654	5.52	0.0812	50	32.717
S3	33.654	5.52	0.126	50	32.717
S4	33.654	5.52	1.33	48.8	32.717
S5	33.654	5.52	0.0539	50	32.717
S6	33.654	5.52	0.116	50	32.717
S7	33.654	5.52	0.234	50	2.714
S8	33.654	5.52	0.131	50	32.314
S9	32.994	5.52	0.154	49.9	32.674

TABLE 8.4

Quantity of Emissions Produced (Pattern Making by 3D Printing)

Scenario: S9 (Pattern Making with 3D Printing Process)

GWP	14.3
AP	0.102
EP	5.2
ODP	399
FAETP	0.0914
HTP	27.1
POPC	5.54

Inputs/Outputs	IN: Sand cast
Flows	100%
Resources	49.8%
Others	
Deposited goods	0.0226%
Emissions to air	0.154%
Emissions to fresh water	49.9%
Emissions to sea water	0.0563%
Emissions to agricultural soil	2.72E-008%
Emissions to industrial soil	1.63E-008%

FIGURE 8.2   Results for GABI model (pattern making by 3D printing).

TABLE 8.5
Detailed Analysis of All Models (Data Used for All Scenarios)

Scenario: S9 (Pattern Making with 3D Printing Process)

URM (Kg/FU)	14.3
UE (Kwh/FU)	0.102
ETA (%)	5.2
ETW (%)	399
SW (Kg/FU)	0.0914

URM: Used raw material for pattern making in 3D printing process; UE: used energy during 3D printing process; ETA: Emissions to air; ETW: Emissions to water; SW: Solid waste

TABLE 8.6
Environmental Impacts by All Different Alternatives

S	GWP	AP	EP	ODP	FAETP	HTP	POPC
S1	20.9	0.126	7	5.31	0.109	51.8	6.9
S2	14.5	0.0527	3.57	10.7	0.0943	49.7	3.43
S3	14.8	0.0537	3.64	134	0.0966	49.8	3.53
S4	14.9	0.08	7.92	4.47	0.117	50.5	7.39
S5	14.5	0.0526	3.56	4.38	0.0939	49.7	3.42
S6	20.7	0.126	6.98	5.2	0.109	51.8	6.84
S7	20.9	0.126	7	5.31	0.109	51.8	6.9
S8	17.3	0.113	6.13	4.78	0.0872	38.6	6.15
S9	14.3	0.102	5.52	399	0.0914	27.1	5.54

Table 8.5 contains the inventory data of all possible scenarios. Reusing waste sand provides a great reduction in solid waste. Also, 3D printing reduces the consumption of raw material. This segment contains the results of each scenario's environmental effect.

Table 8.6 shows the combination of all impacts from different alternatives.

Table 8.7 shows possible combinations of different scenarios. This will help users choose the best possible combination for the sand-casting process. Table 8.7 contains 20 possible combinations.

As shown in the table, the best feasible combination is 3D printing combined with wind or hydropower as a source of electricity. At the moment, it is not possible to generate all of the electricity from wind, but in the future, it may be conceivable. Every country is engaged in research and development in the field of renewable energy sources. Emission factors $KgCO_2$ per unit has been lowered from 20.9 to 4.1, and $KgSO_2$ per unit has been reduced from 0.126 to 0.0156. This chapter helps us to find the best available alternative for the sand-casting process, because this chapter provides emission details for each model.

TABLE 8.7
Result by Combining All Cases

CB	A	B	C	D	E	F	G	H	Kg CO_2 eq.	Kg SO_2 eq
1	✓	✗	✗	✗	✓	✗	✗	✗	14.3	0.0527
2	✓	✗	✗	✗	✗	✓	✗	✗	14.5	0.0527
3	✓	✗	✗	✗	✗	✗	✓	✗	10.9	0.0397
4	✗	✓	✗	✗	✓	✗	✗	✗	14.6	0.0537
5	✗	✓	✗	✗	✗	✓	✗	✗	14.8	0.0537
6	✗	✓	✗	✗	✗	✗	✓	✗	11.2	0.0407
7	✗	✗	✓	✗	✓	✗	✗	✗	14.7	0.0800
8	✗	✗	✓	✗	✗	✓	✗	✗	14.9	0.0800
9	✗	✗	✓	✗	✗	✗	✓	✗	11.3	0.0670
10	✗	✗	✗	✓	✓	✗	✗	✗	14.3	0.0526
11	✗	✗	✗	✓	✗	✓	✗	✗	14.5	0.0526
12	✗	✗	✗	✓	✗	✗	✓	✗	10.9	0.0396
13	✓	✗	✗	✗	✓	✓	✓	✗	10.7	0.0397
14	✗	✗	✗	✗	✓	✓	✓	✗	11	0.0407
15	✗	✗	✗	✗	✓	✓	✓	✗	1.1	0.0670
16	✗	✗	✗	✓	✓	✓	✓	✗	10.7	0.0396
17	✓	✗	✗	✗	✓	✓	✓	✓	4.1	0.0157
18	✗	✓	✗	✗	✓	✓	✓	✓	4.4	0.0167
19	✗	✗	✓	✓	✓	✓	✓	✓	4.5	0.0430
20	✗	✗	✗	✓	✓	✓	✓	✓	4.1	0.0156

CB: Combination; A: Wind energy; B: Solar energy; C: Energy by biomass; D: Energy by hydropower;
E: Resin reduction; F: Sand recycling; G: Recycling of scraps; H: 3D printing

8.4.3 OPPORTUNITIES FOR ADDITIVE MANUFACTURING IN INDUSTRY 4.0 AND SUSTAINABILITY

The modern industrial revolution is related to the digital world, and AM is a part of it. It denotes a completely new method of manufacturing that involves layering materials before the desired result is achieved. Because of Industry 4.0, there has been a noteworthy development in smart infrastructure, which may impose large environmental burdens; yet, standard LCA methodologies are often unable or insufficient to quantify such consequences. As a result, a gap has developed between the field of environmental assessment and evident advances in the manufacturing sphere. A trade-off is frequently made between resource efficiency and the possible benefits of new technology in terms of material reduction and other negative implications, such as increased energy usage.

8.5 CONCLUSION

This chapter presents the environmental impacts during the manufacturing phase and life cycle inventory of the sand-casting process. LCA methodology is used for the assessment of environmental impact during the sand-casting process. The proposed methodology has been successfully used in developing a generalized approach needed for predefined decision-making in sustainable manufacturing processes. The use of electricity (wind or hydropower) as a source of energy will reduce carbon dioxide emissions by approximately 30.62%. The energy used, solid waste, and harmful air emission contaminants are all major environmental issues in the conventional process. Solid waste in landfills has the biggest environmental effect. Solid waste recycling will result in a 91% cut in solid waste. 3D printing alone reduces carbon dioxide emissions by 31%. The best possible case can be achieved by combining different scenarios. Table 8.7 compares the base scenario with the best possible scenario. AM is a component of the new industrial revolution, which is inextricably tied to the digital realm. It's a brand-new manufacturing method that involves layering ingredients until the desired result is achieved. In the long run, this new manufacturing model will allow us to eliminate a lot of manufacturing tools and adapt production processes to market demands with a lot more flexibility. In many ways, recent technological advancements indicate that AM is on the verge of becoming a viable alternative to conventional production procedures.

8.6 FUTURE SCOPE

Because of the use of AM technology and the fact that many of the new business models are still in their early stages, more research is required. The AM process has developed to the point where it is now able to recreate complicated and functional mechanical equipment thanks to many technological developments. AM technologies were too slow, ineffectual, and expensive compared to conventional process. A cradle-to-gate LCA of both processes studied here is planned for future work.

REFERENCES

(1) Peng, T., Kellens, K., Tang, R., Chen, C., & Chen, G. (2018). Sustainability of additive manufacturing: An overview on its energy demand and environmental impact. *Additive Manufacturing,* 21, 694–704.
(2) Ford, S., & Despeisse, M. (2016). Additive manufacturing and sustainability: An exploratory study of the advantages and challenges. *Journal of Cleaner Production*, 137, 1573–1587.
(3) Sreenivasan, R., Goel, A., & Bourell, D. L. (2010). Sustainability issues in laser-based additive manufacturing. *Physics Procedia,* 5, 81–90.
(4) Mani, M., Lyons, K. W., & Gupta, S. K. (2014). Sustainability characterization for additive manufacturing. *Journal of Research of the National Institute of Standards and Technology*, 119, 419.

(5) Taddese, G., Durieux, S., & Duc, E. (2020). Sustainability performance indicators for additive manufacturing: A literature review based on product life cycle studies. *The International Journal of Advanced Manufacturing Technology*, 107(7), 3109–3134.

(6) Godina, R., Ribeiro, I., Matos, F., T Ferreira, B., Carvalho, H., & Peças, P. (2020). Impact assessment of additive manufacturing on sustainable business models in Industry 4.0 context. *Sustainability*, 12(17), 7066.

(7) Fredriksson, C. (2019). Sustainability of metal powder additive manufacturing. *Procedia Manufacturing*, 33, 139–144.

(8) Angioletti, C. M., Sisca, F. G., F G, F., Luglietti, R., Taisch, M., & Rocca, R. (2016). Additive manufacturing as an opportunity for supporting sustainability through the implementation of circular economies. In *21st Summer School Francesco Turco 2016* (pp. 25–25). AIDI-Italian Association of Industrial Operations Professors.

(9) Diegel, O., Kristav, P., Motte, D., & Kianian, B. (2016). Additive manufacturing and its effect on sustainable design. In *Handbook of Sustainability in Additive Manufacturing* (pp. 73–99). Springer, Singapore.

(10) Machado, C. G., Despeisse, M., Winroth, M., & da Silva, E. H. D. R. (2019). Additive manufacturing from the sustainability perspective: Proposal for a self-assessment tool. *Procedia CIRP*, 81, 482–487.

(11) Niaki, M. K., Torabi, S. A., & Nonino, F. (2019). Why manufacturers adopt additive manufacturing technologies: The role of sustainability. *Journal of Cleaner Production*, 222, 381–392.

(12) Floriane, L., Enrico, B., Frédéric, S., Nicolas, P., & Paolo, C. (2018, July). TEAM: A tool for eco additive manufacturing to optimize environmental impact in early design stages. In *IFIP International Conference on Product Lifecycle Management* (pp. 736–746). Springer, Cham.

(13) Yosofi, M., Kerbrat, O., & Mognol, P. (2018). Energy and material flow modelling of additive manufacturing processes. *Virtual and Physical Prototyping*, 13(2), 83–96.

(14) Suárez, L., & Domínguez, M. (2020). Sustainability and environmental impact of fused deposition modelling (FDM) technologies. *The International Journal of Advanced Manufacturing Technology*, 106(3), 1267–1279.

(15) Jamwal, A., Agrawal, R., & Sharma, M. (2021). Life cycle engineering: Past, present, and future. In *Sustainable Manufacturing* (pp. 313–338). Elsevier, Netherlands.

(16) Villamil, C., Nylander, J., Hallstedt, S. I., Schulte, J., & Watz, M. (2018). Additive manufacturing from a strategic sustainability perspective. In *DS 92: Proceedings of the DESIGN 2018 15th International Design Conference* (pp. 1381–1392). Design Society, Glasgow.

(17) Khalid, M., & Peng, Q. (2021). Sustainability and environmental impact of additive manufacturing: A literature review. *Computer-Aided Design and Applications*, 18(6), 1210–1232.

(18) Kohtala, C. (2015). Addressing sustainability in research on distributed production: An integrated literature review. *Journal of Cleaner Production*, 106, 654–668.

(19) Burkhart, M., & Aurich, J. C. (2015). Framework to predict the environmental impact of additive manufacturing in the life cycle of a commercial vehicle. *Procedia CIRP*, 29, 408–413.

(20) Yadav, A., Jamwal, A., Agrawal, R., & Kumar, A. (2021). Environmental impacts assessment during sand casting of Aluminium LM04 product: A case of Indian manufacturing industry. *Procedia CIRP*, 98, 181–186.

(21) Al-Meslemi, Y., Anwer, N., & Mathieu, L. (2018). Environmental performance and key characteristics in additive manufacturing: A literature review. *Procedia CIRP*, 69, 148–153.

(22) Kai, D. A., de LIMA, E. P., Cunico, M. W. M., & da Costa, S. E. G. (2016, October). Measure additive manufacturing for sustainable manufacturing. In *ISPE TE* (pp. 186–195). IOS Press, Amsterdam.

(23) Laverne, F., Marquardt, R., Segonds, F., Koutiri, I., & Perry, N. (2019). Improving resources consumption of additive manufacturing use during early design stages: A case study. *International Journal of Sustainable Engineering*, 12(6), 365–375.

(24) Mançanares, C. G., de Senzi Zancul, E., & Miguel, P. A. C. (2015). Sustainable manufacturing strategies: A literature review on additive manufacturing approach. *Product: Management and Development,* 13(1), 47–56.

(25) Oros Daraban, A. E., Negrea, C. S., Artimon, F. G., Angelescu, D., Popan, G., Gheorghe, S. I., & Gheorghe, M. (2019). A deep look at metal additive manufacturing recycling and use tools for sustainability performance. *Sustainability*, 11(19), 5494.

(26) Gupta, H., Kumar, A., & Wasan, P. (2021). Industry 4.0, cleaner production and circular economy: An integrative framework for evaluating ethical and sustainable business performance of manufacturing organizations. *Journal of Cleaner Production*, 295, 126253

9 Build Orientation Optimization of Car Hoodvent with Additive Manufacturing

*Marina A. Matos, Ana Maria A.C. Rocha,
Lino A. Costa and Ana I. Pereira*

CONTENTS

DOI: 10.1201/9781003123866-9

9.1   INTRODUCTION

Rapid prototyping (RP), a technology for manufacturing models in less time, has increased over the years. Many models have been implemented in manufacturing companies due to its effectiveness in reducing the overall product development cost and time (Canellidis et al., 2006). RP is often applied in software engineering and applications in several areas, such as aerospace, automotive, product development, and healthcare. Figure 9.1 shows the percentage of areas where rapid prototyping is used.

RP can be divided into two techniques: additive manufacturing and subtractive manufacturing. Additive manufacturing (AM) or 3D printing includes a set of technologies for manufacturing 3D objects by overlapping material layer by layer, starting from scratch. RP technology was introduced by Kruth (1991) in the 1980s. It starts from a 3D mesh model produced in computer-aided design (CAD) software, followed by converting the model into a standard tessellation language (STL). Then, the STL file is sent to a 3D printer for building the part. Subtractive manufacturing is based on removing material from a block of raw materials.[2] These two types of manufacture can be viewed in Figure 9.2.

Additive manufacturing has been widely used in recent years by several companies producing 3D objects, which are defined by the addition of layer-by-layer material, using different construction materials, such as resin, plastic, glass, rubber, ceramics, concrete and metal (Abdulhameed et al., 2019; Ngo et al., 2018; Zhang et al., 2016). Currently, AM is applied in different areas, such as medical sciences (for example, dental restorations and medical implants), jewelry, the footwear industry, the automotive industry and the aviation industry (Bikas et al., 2016; Del Re et al., 2019). Using this technology, it is possible to build very complex geometric parts without requiring many post-processing procedures. Moreover, this process involves the reduction of material waste and lower transport and storage costs and can produce large quantities of 3D models faster than other traditional methods (Jiang et al., 2018).

Over the last decades, several processes have been included in AM, such as fused deposition modeling (FDM), stereolithography, selective laser fusion, selective laser sintering, laser cladding, laminated object manufacturing and laser vapor deposition,

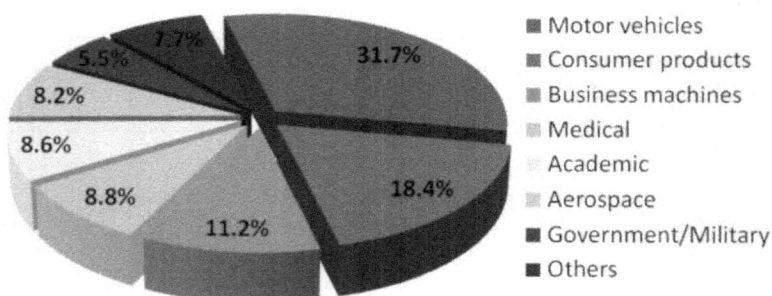

FIGURE 9.1   Rapid prototyping worldwide.[1]

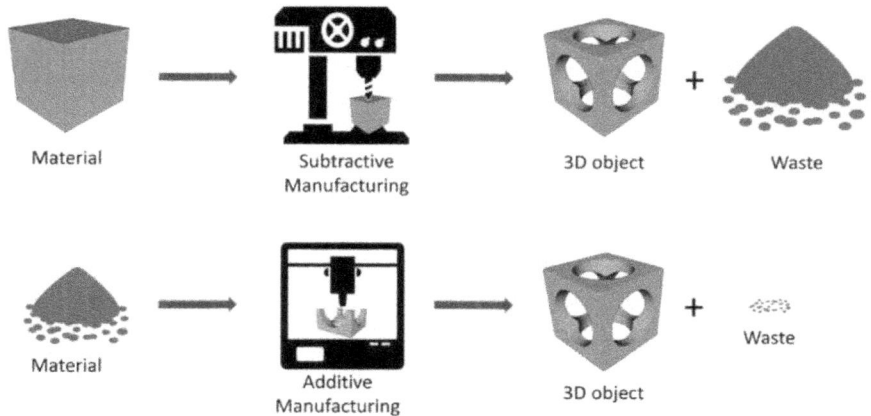

FIGURE 9.2   Additive and subtractive manufacturing.

among others (Lee et al., 2017; Shanmugam et al., 2021; Taufik and Jain, 2013). FDM is one of the most-used processes for the production of 3D parts and will be the technique discussed in this chapter. FDM is a process that creates a 3D object from a thermoplastic filament, which is heated to its melting point and then extruded layer by layer. This process has as its main advantages the fact that it is flexible, allows the development of 3D parts in different orientations (Cerneels et al., 2013), improves resource efficiency, gives greater durability to the model and manages to produce parts on a small scale and with complex geometry (Ford and Despeisse, 2016). In addition, this technology generates less material waste and requires less time and reduced costs during the production of 3D objects (Qin et al., 2019; Sheoran and Kumar, 2020).

AM performance relies on the orientation of the parts in the printer platform, where each part should have an optimal orientation to improve the surface quality in order to reduce the building time and finally decrease the total costs (Rocha et al., 2018). The best build orientation is a crucial factor since it affects several factors such as the amount of material, the printing time, the material cost, the quality of the final part (roughness) and the amount of support material. In addition, an optimal orientation makes it possible to reduce or eliminate errors involved throughout the 3D model printing process (Matos et al., 2020b; Zhang et al., 2016).

There are different approaches to obtain the best build orientation of a 3D CAD model based on single-objective optimization approaches (Canellidis et al., 2009; Masood et al., 2003; Matos et al., 2019b; Phatak and Pande, 2012). In Thrimurthulu et al. (2004) and Canellidis et al. (2006), methodologies addressing optimal part build orientation taking into account the surface quality, the surface roughness and the build deposition time are shown. In the work of Rocha et al. (2018), an optimization approach to solve the 3D part construction orientation problem was introduced. The quality measures considered in this work were: the staircase effect, the support area and the build time. For this study, two global optimization methods were applied, the electromagnetism-like and the stretched simulated annealing algorithms, and four

3D models were studied. The obtained best build orientations for the different parts showed that both algorithms presented good results. Matos et al. (2020) used the electromagnetism-like algorithm to determine the best orientation of the six 3D CAD models. Six quality measures, the volumetric error, the support area, the staircase effect, the build time, the surface roughness and the surface quality, were optimized separately, and several optimal solutions were obtained. The experiments validate the proposed approach.

In recent years, different multi-objective approaches have also been proposed in order to determine the optimal orientation in the building of 3D objects (Brika et al., 2017; Gurrala and Regalla, 2014; Jibin, 2005; Matos et al., 2021). Li et al. (2010) proposed a multi-objective optimization approach focused on obtaining the desired orientations based on the following objectives: the support area, the build time and the surface roughness. For this, a profile extent function combined with the particle swarm optimization method was used. Padhye and Deb (2011) developed a multi-objective approach considering the surface roughness and the build time as objective functions for different models. The non-dominated sorting genetic algorithm (NSGA-II) and multi-objective particle swarm optimization (MOPSO) algorithms were used to obtain the Pareto front. A combination of three approaches, metamodeling, NSGA-II and technique for order of preference by similarity to ideal solution (TOPSIS), was applied in Khodaygan and Golmohammadi (2018) in order to obtain the best build orientations in 3D objects. Thus, the measures studied in this work were the roughness of the piece and its printing time in order to minimize its values. The authors compared and discussed the results of each method. In the study carried out by Qin et al. (2019), a method based on diffuse multi-attribute decision-making to determine the optimal 3D part build orientation is presented. They concluded that this method does not present a demanding high computational cost when compared to other methods used for this type of problem. Matos et al. (2019a) applied the weighted Tchebycheff scalarization method embedded in the EM optimization algorithm. A set of optimal solutions were identified, resulting in a low printing time and low necessary use of material. NSGA-II and generalized differential evolution 3 (GDE3) were used in Mele and Campana (2020) in order to increase the sustainability of additive manufacturing processes for the construction of a part in a given orientation. In this study, the authors were able to verify that the GDE3 algorithm obtains more beneficial solutions compared with the NSGA-II algorithm. Olsen and Kim (2020) developed a study to optimize part topology and build orientation simultaneously. The presented methodology allowed reduction of the material support when printing 3D parts, with the ability to obtain models with high quality. Matos et al. (2021a) applied a many-objective approach for a 3D CAD model, where NSGA-II was used with four quality measures (the support area, the build time, the surface roughness and the surface quality). In the first approach, a bi-objective optimization analysis was performed for each pair of objectives, where optimal solutions were found. Then, an optimization based on many objectives was applied in order to optimize the four measures simultaneously. The authors of this work obtained several optimal build orientations. In a paper by Sheng et al. (2020), two approaches to calculate optimal build orientations, single-objective and multi-objective optimization, were applied. Initially, they optimized the support structures using the particle swarm optimization (PSO) algorithm. Then, a multi-objective approach to optimize the support structures and the build time at the same time was studied using a MOPSO algorithm. They concluded that both algorithms found several optimal build orientations. An article by Di Angelo et al. (2020) presents a multi-objective formulation for different 3D models. Two objective functions, surface quality and the total cost of parts, were used, according to the build orientation of the models. The main goal of the paper was to minimize the build orientation problem in order to obtain optimal orientations. Thus, the Pareto fronts were obtained using the multi-objective S-metric evolutionary selection algorithm. In the work developed in Matos et al. (2021b), the best printing orientations for a 3D CAD part, considering the support area, the staircase effect and the build time, were found. In this study, the multi-objective NSGA-II was applied in order to optimize two approaches, bi-objective optimization and multi-objective optimization.

The main goal of this work focuses on the optimization of the build orientation problem of the car hoodvent model with the purpose of providing the decision-maker the optimal build orientations and their trade-offs. Three measures will be considered: the total contact area of the external support with the object (herein denoted the support area), the time required to build the object (herein denoted the build time) and the surface roughness. First, a single-objective approach is developed to find the optimal build orientations of the three measures mentioned previously but optimized separately. Then, a bi-objective approach is carried out to optimize the combinations of two measures. Finally, a many-objective approach optimizing the three measures simultaneously is performed. The genetic algorithm (GA) and NSGA-II methods implemented in the MATLAB Global Optimization Toolbox will be used through the *ga* function for solving the single-objective problem, and the *gamultiobj* function will be used for solving the bi-objective and many-objective problems.

This chapter is organized as follows. Section 9.2 introduces the part build orientation problem, presenting the orientation of 3D CAD parts, the phenomenon of the staircase effect that occurs when printing the parts and the orientation computational system. The model for the experiments is presented in Section 9.3. The three quality measures that will be optimized are described in Section 9.4. The single-objective optimization approach is presented in Section 9.5, as well as the obtained results. The experiments based on bi- and multi-objective optimization are presented in Sections 9.6 and 9.7. The numerical experiments are discussed and compared to each other in Section 9.8, and Section 9.9 contains the conclusions of this chapter and some recommendations for future work.

9.2   PART BUILD ORIENTATION

In this section of the chapter, the issue of part orientation will be addressed. Here, a brief introduction related to the orientation of 3D CAD parts is given, followed by the concepts associated with the staircase effect, and the orientation computational process in order to obtain optimal orientations in the construction of three-dimensional models is also mentioned.

9.2.1 INTRODUCTION

The selection of the best orientation is a very important factor in additive manufacturing, since it contributes to the reduction of the manufacturing time, reduces the amount of support structure, gives a better precision, reduces the construction cost of the part and allows a better surface finish (Das et al., 2017; Pandey et al., 2007). In order to find the optimal orientation of a 3D object, it is necessary to rotate it on different axes (x, y and z axes). Figure 9.3 shows the rotations around the different axes. These rotations follow the rule of the right hand (counterclockwise, that is, direction of positive rotation).

The quality measures involved in this study and presented in Section 9.4 require a vectorial direction $= (x, y, z)^T$, which is calculated using Equation (1):

$$x^2 + y^2 + z^2 = 1 \qquad (1)$$

where the variables x, y, and z are given by:

$$\begin{cases} x = \sin(\beta)\cos(\rho) \\ y = \sin(\beta)\sin(\rho) \\ z = \cos(\beta) \end{cases}$$

In Figure 9.4, the direction unit vector with all the variables represented previously can be seen.

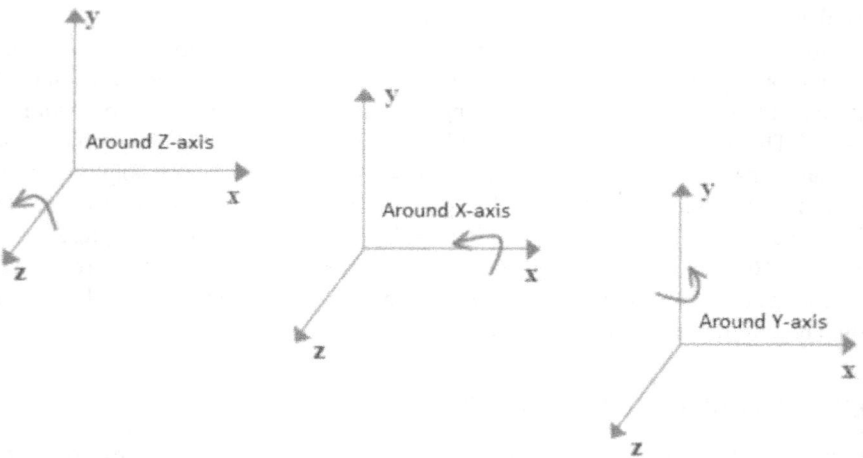

FIGURE 9.3   Positive rotation around axes.

Source: Nakano and Cunha, 2007

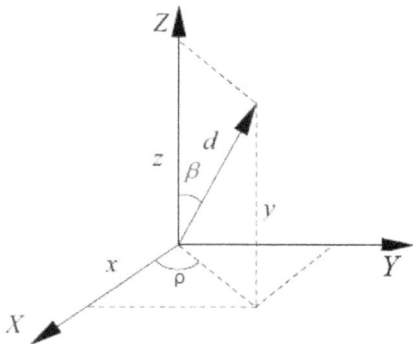

FIGURE 9.4   Unit vector of build orientation.

Source: Jibin, 2005

9.2.2   STAIRCASE EFFECT

In this chapter, the main objective is to optimize the build orientation of the 3D part in order to reduce its surface roughness, build time and support area, taking into account several factors such as the staircase effect. The staircase effect may emerge when the thickness of the layers is uniform on curved surfaces, when there is shrinking of the layers and when the laser angle of the printer is not correct. This effect occurs when the deposition of material is done from the bottom up, with the shape of the lower levels altering the upper layers. This effect can cause dimensional errors and make the part rough on its surface (Raju et al., 2014). In other words, the smaller the layer thickness, the smaller the staircase effect and the greater the surface quality of the model. A method that optimizes the cut layer thickness is presented by Wang et al. (2015). The aim of this paper is to save time in the printing of a 3D model and consequently to obtain a high surface quality. For this purpose, the authors decided to divide the model into subparts in order to optimize the slicing of each part separately. This approach reduced the printing time of the several parts studied by 30–40% and obtained objects with a better surface resolution.

Figure 9.5 shows the phenomenon of the staircase effect of a model. The error associated with the staircase effect occurs due to the layer thickness and the slope of the part surface (Jaiswal et al., 2018).

The maximum deviation between the part surface and the printed object caused by the staircase effect is denoted as the cusp height (CH), which is computed by the maximum deviation from the layered part to the CAD surface measured in the normal direction to the CAD surface. The cusp height is related to the angle α formed by the slicing direction d and the model surface normal n and to the layer thickness t. The cusp height is given by $CH = t\cos(\alpha)$, where $\cos(\alpha) = |d \cdot n|$. Larger values for cusp height are produced by thicker layers and/or higher values of $\cos(\alpha)$, and consequently a more inaccurate surface will be detected (Alexander et al., 1998; Pereira et al., 2018).

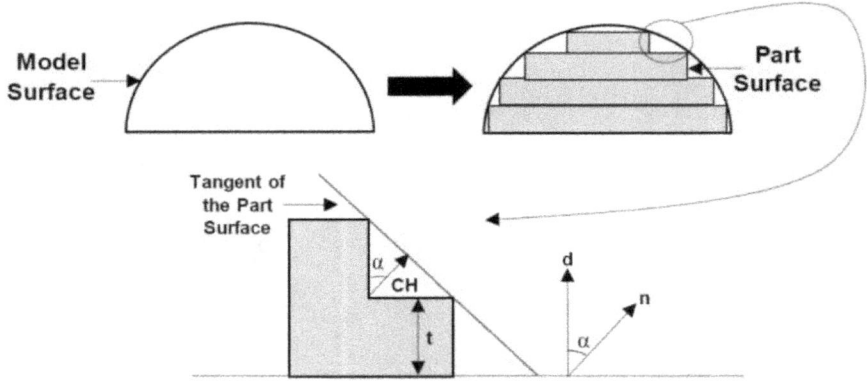

FIGURE 9.5   Staircase phenomenon.

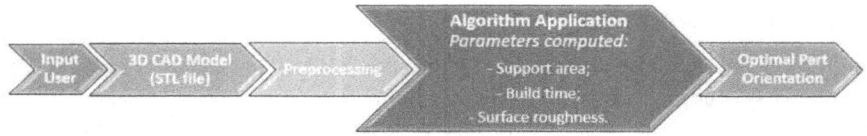

FIGURE 9.6   Computational process to obtain the best orientation of a 3D CAD model.

9.2.3 Orientation Computational System

The process of finding the best orientation for a 3D model starts with a 3D mesh model that can be created by structures built in CAD software and then converted to a STL file (see Section 9.3). Then, pre-processing is carried out, which consists of inserting the STL file in a 3D printing program and defining all parameters for printing the 3D part. In the following, the optimization problem considering the objective functions in the process is solved by an optimization algorithm. Finally, the optimal orientations are obtained for printing the model. In Figure 9.6, the entire digital process can be seen before making the model on the 3D printer.

9.3 CAR HOODVENT MODEL

The model used in this study is denoted by car hoodvent, which is a front ventilation grill for a car, as shown in Figure 9.7. The size of this model is $389.5 \times 147.5 \times 18.5 \left(\text{width} \times \text{height} \times \text{depth} \right)$ in mm and the volume is 144.2 cm^3. The layer thickness used in this work was 0.2 mm.

The STL file is a description of the geometric characteristics of the 3D model by an approximation (tessellation) of the CAD model. Hence, the representation of the model is done by a mesh of triangles, where only the surface geometry of a three-dimensional object without any representation of color, texture or other common

FIGURE 9.7   3D CAD representation of the car hoodvent model.

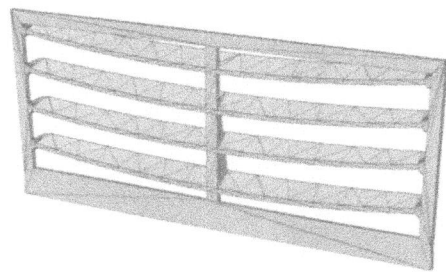

FIGURE 9.8   Triangular representation of the car hoodvent model.

attributes of the CAD model are described. The car hoodvent is defined by 8646 triangles, and its representation is presented in Figure 9.8.

9.4   QUALITY MEASURES

This section presents the concepts of the three objective functions studied in this work: the support area, the build time and the surface roughness. For all of them, the mathematical formulation and the landscape of each function of the car hoodvent model are presented.

9.4.1   SUPPORT AREA

The amount of support affects the building time of the part as well as the surface accuracy. This can be measured by the support area (SA) or support volume. The support volume has an effect on building the model, and computationally, this is very complex to calculate. It is the volume of the region that is between the platform of the 3D printer and the layer under construction. The support area is the measure that accounts for the amount of support to be used in the construction of the part and is defined as the total area of the downward-facing facets. This is measured through the total contact area of the external support with the object and consequently affects the building time of the part and also the final cost. Support area mainly influences

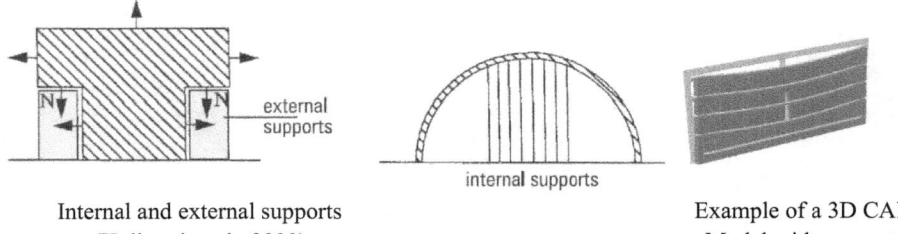

Internal and external supports   Example of a 3D CAD
(Kulkarni et al., 2000)   Model with supports

FIGURE 9.9 Support in 3D models.

post-processing and surface finish. This is computationally simpler and also the most important when it comes to part accuracy (Jibin, 2005; Matos et al., 2020).

The SA objective function is expressed by:

$$SA = \sum_i A_i \left| d \cdot n_i \right| \delta \tag{2}$$

$$W_{ij} = \begin{cases} W'_{ij} & \text{if } W'_{ij} > treconf \\ treconf, & \text{otherwise} \end{cases}$$

where A_i is the area of each triangular facet i, n_i is the normal unit vector of each triangular facet i, d is the unit vector of the direction of construction and δ is the initial function (Matos et al., 2021b).

In Figure 9.9, two examples of 3D objects that need support during their printing are presented.

Figure 9.10 depicts the SA objective function landscape for the car hoodvent model. From the figure, it can be seen that this objective function is nonconvex with multiple local optima.

9.4.2 BUILD TIME

The time required to create an object accurately is composed of the time of the 3D printer to deposit the material layer by layer plus the time for support removal and surface finishing (Canellidis et al., 2006). According to Jibin (2005), the build time is the precise time for the platform to move downward during the construction of each layer and depends on the total number of slices of the solid. In addition, the number of slices depends on the height of the build orientation of the object, so the build time is proportional to the height of the model. Therefore, minimizing the number and height of layers makes it possible to shorten the build time of the part.

The build time (BT) is given by

$$BT = \max_i \left(d^T v_i^1, d^T v_i^2, d^T v_i^3 \right) - \min_i \left(d^T v_i^1, d^T v_i^2, d^T v_i^3 \right) \tag{3}$$

SA

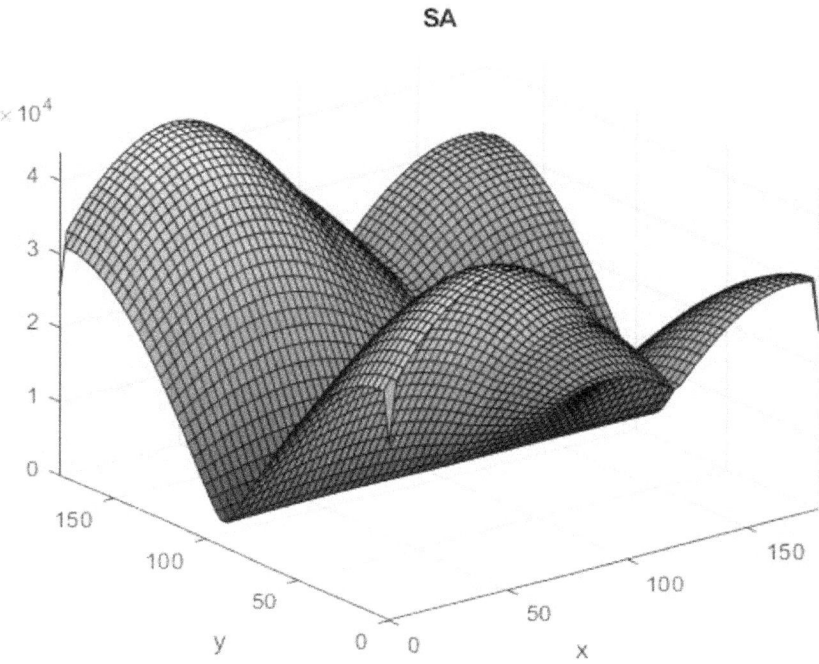

FIGURE 9.10   *SA* objective function landscape of car hoodvent.

where d is the direction vector and v_i^1, v_i^2, v_i^3 are the vertex triangle facets i (Matos et al., 2021a).

The objective function BT for the part car hoodvent is represented in Figure 9.11. It appears that this measure is non-convex and has multiple local optima.

9.4.3 SURFACE ROUGHNESS

Several factors affect the surface roughness (RA), such as the layer thickness, the number of support structures and the orientation of the part (Jaiswal et al., 2018; Zhang et al., 2017). In addition, it may imply less resistance of the part during the work performance, leading to poor product quality (Gok, 2015). Over the past few years, several authors have studied different approaches to define surface roughness (Ahn et al., 2008; Byun and Lee, 2006; Campbell et al., 2002; Canellidis et al., 2009; Mason, 2007; Mohan Pandey et al., 2003; Pandey et al., 2003). For a better precision of the value of RA , Behnam (2011) performed a review of the various equations of surface roughness, forming a set of equations that calculate the value of RA depending on the orientation angle.

The surface roughness for each triangle facet i , RA_i , is defined taking into consideration the build angle θ and is given by

BT

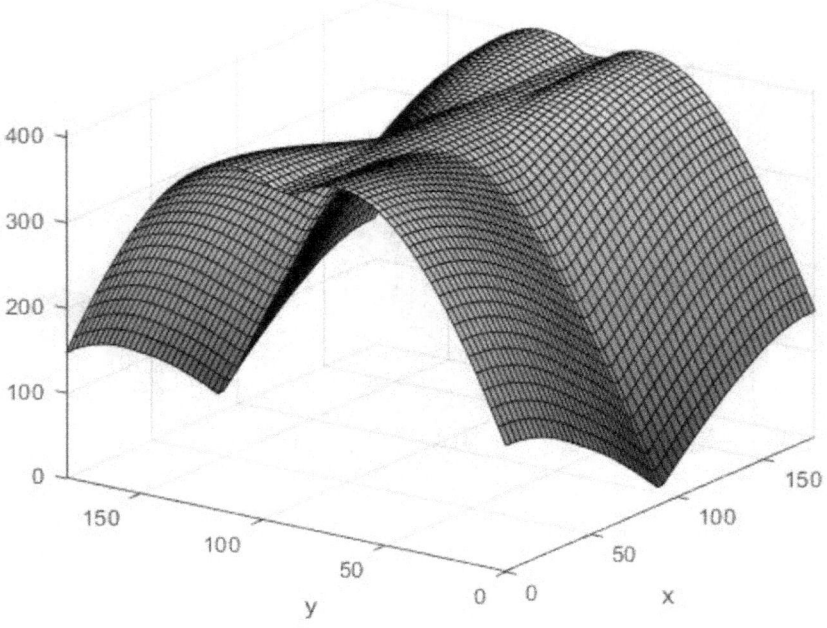

FIGURE 9.11 *BT* objective function landscape of car hoodvent.

$$
\begin{cases}
70.82 \dfrac{t}{\cos(\theta_i)}, & if\, 0° \le \theta_i \le 70° \\[2ex]
\dfrac{1}{20}\left(90RA_i^{70} - 70RA_i^{90} + \theta_i\left(RA_i^{90} - RA_i^{70}\right)\right), & if\, 70° < \theta_i < 90° \\[2ex]
117.6t, & if\, \theta_i = 90° \\[2ex]
RA_i^{\theta_i - 90}\left(1 + w\right), & if\, 90° < \theta_i \le 135° \\[2ex]
\dfrac{1000}{2} t \left| \dfrac{\cos(90 - \theta_i) - \phi}{\cos(\phi)} \right|, & if\, 135° < \theta_i \le 180°
\end{cases}
\qquad (4)
$$

where t is the thickness of the layer; $\theta_i = 90 - \alpha_i$; α_i is the angle of the unit vector of the direction and the normal unit vector for each triangle facet i; and RA_i^{70} and RA_i^{90} are the values of RA_i when $\theta_i = 70°$ and $\theta_i = 90°$, respectively. w is a dimensionless adjustment parameter for supported facets, and ϕ is a phase shift in the range of $5° \le \phi \le 15°$ depending on the layer thickness (Behnam, 2011). The value 70.82 in the first branch of Equation (4) refers to a value inside the confidence interval $(69.28 \sim 72.36)$ used in Pandey (2010); $w=0.2$ as proposed in Pandey et al. (2003), and $\phi = 5°$ as in Behnam (2011).

RA

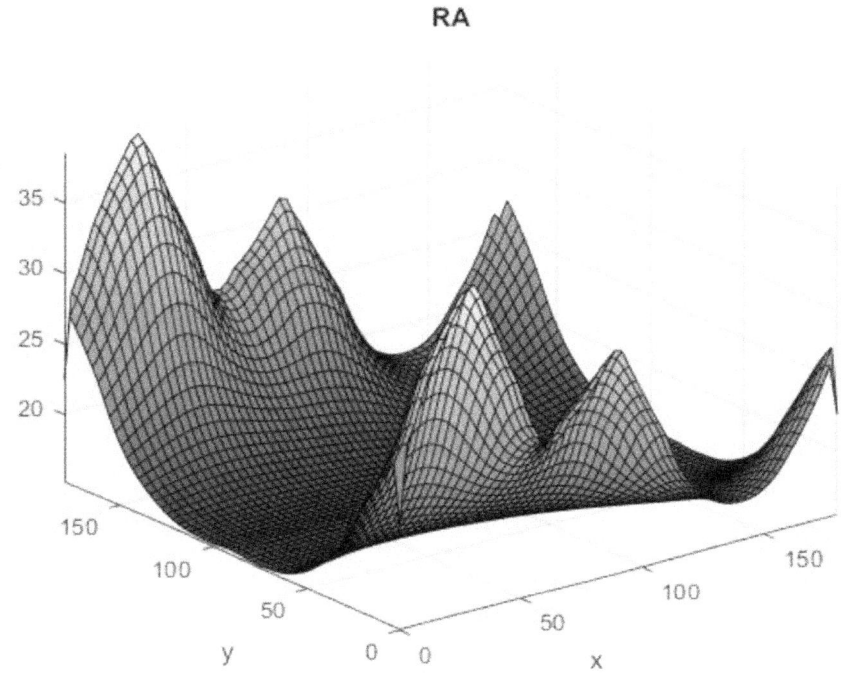

FIGURE 9.12 *RA* objective function landscape of car hoodvent.

The average surface roughness, considering the area of the triangular facets, can be calculated as

$$RA = \frac{\sum_i \left(RA_i A_i \right)}{A_i} \qquad (5)$$

where RA_i is the roughness (in μm) of each triangular surface i, and A_i is the area of triangular facet i. The smaller the RA values, the smoother the surface.

The landscape of the objective function RA is represented in Figure 9.12 for the car hoodvent model. Again, it turns out that this objective function has multiple local optima and is non-convex.

9.5 SINGLE-OBJECTIVE OPTIMIZATION APPROACH

In this section, a single-objective approach of this work is presented. First, the mathematical formulation of the problem is described, followed by the methodology used. Finally, the optimal building orientation results are shown.

9.5.1 PROBLEM

The single-objective build orientation problem of a 3D CAD model only considers the rotation angles θ_x of the model around the x-axis and θ_y around y-axis, since the 3D printer's printing platform is fixed to the z-axis, not affecting the part construction process. The mathematical formulation of the single-objective optimization problem is given by

$$min f\left(\theta_x, \theta_y\right)$$

$$CO2 = \sum_{i=1}^{n}\sum_{j=1}^{n_i} e_{ij}^k \left(t(\tau), R_\tau\right), \forall k \tag{6}$$

$$J_1 = \left\{O_{11}, O_{12}, O_{13}\right\}$$

where f is the objective function, and θ_x and θ_y are the rotation angles along the x-axis and the y-axis, respectively, where each angle is between $0°$ and $180°$.

This initial study aims to find the best build orientation taking into account each of the measures (SA, BT and RA) presented in the previous section, optimizing the single-objective problem [Equation (6)].

9.5.2 METHODOLOGY

The ga function from the MATLAB Global Optimization Toolbox (MATLAB, 2019) is used to obtain the optimal build orientation for the car hoodvent model studied in this chapter. The genetic algorithm, a population-based stochastic algorithm that mimics biological evolution, is implemented in this function. In this algorithm, the best individuals in the current population are likely to be selected to produce offspring for next generations by mutation and crossover. The population evolves towards an approximation of the optimal solution over successive generations (Goldberg, 1989). The genetic algorithm is implemented in MATLAB software through the ga predefined function. The default values of the ga function were used. The size of the population was set to 50 individuals, the stopping criterion was a maximum number of generations of 200 and 30 independent runs were considered. The direction $d = (0,0,1)^T$ was defined as the slicing direction after a rotation along $\theta = \left(\theta_x, \theta_y\right)$ angles. This vector is a normalized vector (i.e. $\|d\| = 1$). The numerical experiments were performed in a PC Intel Core(TM) i7–7500U CPU with 2.9 GHz and 12.0 GB of memory RAM. The GA algorithm was coded in MATLAB Version 9.6.0.1472908 (R2019a) Update 9.

9.5.3 RESULTS

Five solutions (A, B, C, D and E) were obtained for the single-objective problem. Each objective function (SA, BT and RA) was separately optimized. The results are presented in Table 9.1.

TABLE 9.1
Optimal Solutions for the Single-Objective Problem

	θ_x	θ_y	SA	BT	RA
A	$\forall\theta_x$	90.00	2.452647×10^3	———	———
B	90.00	180.00	———	1.851336×10^1	———
C	90.00	0.00	———	1.851335×10^1	———
D	134.31	115.70	———	———	1.515983×10^1
E	134.31	64.29	———	———	1.515999×10^1

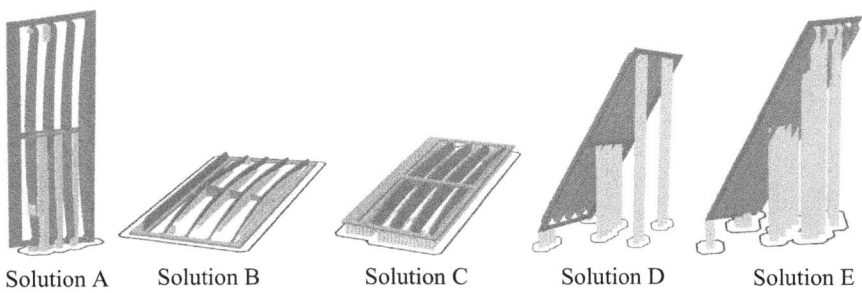

Solution A   Solution B   Solution C   Solution D   Solution E

FIGURE 9.13 Representation of the optimal solutions A to E for the car hoodvent model.

For the *SA* objective function, the optimal solution obtained was $(\forall\theta_x, 90.00)$, with a value of *SA* of 2.452647×10^3, in which the orientation found can be seen in Figure 9.13.

The optimal orientations found for the *BT* objective function were very close: $(90.00, 180.00)$ and $(90.00.0.00)$, where the value of *BT* is 1.851336×10^1 and 1.851335×10^1, respectively. For the surface roughness, the optimal solutions obtained were $(134.31, 115.70)$ and $(134.31, 64.29)$, with values of *RA* similar to 1.515983×10^1 and 1.515999×10^1, respectively. The orientations A, B, C, D and E can be seen in Figure 9.13.

9.6 BI-OBJECTIVE OPTIMIZATION APPROACH

The formulation of the bi-objective optimization problem addressed in this chapter, the methodology used and the obtained results are presented in this section.

9.6.1 Problem

The mathematical formulation of the bi-objective optimization problem is given by

$$\min\left\{f_1\left(\theta_x,\theta_y\right),f_2\left(\theta_x,\theta_y\right)\right\}$$

$$O_{11}=\left\{M_1;M_2\right\},O_{12}=\left\{M_1;M_2\right\},O_{13}=\left\{M_2\right\} \tag{7}$$

$$O_{21}=\left\{M_1;M_2\right\},O_{22}=\left\{M_1;M_2\right\},O_{23}=\left\{M_2\right\}$$

where f_1 and f_2 are the objective functions.

In this section, pairs of the three objectives (SA , BT and RA) will be optimized. Thus, the bi-objective optimization problem will be solved for all possible pair combinations of the three objectives (SA , BT and RA):

o $SA\,vs.\,BT$ —problem (7) with f_1 = SA and f_2 = BT ;
o $SA\,vs.\,RA$ —problem (7) with f_1 = SA and f_2 = RA ;
o $BT\,vs.\,RA$ —problem (7) with f_1 = BT and f_2 = RA .

9.6.2 Methodology

In this work, the elitist NSGA-II is applied (Deb, 2001) to solve problem (7). This algorithm mimics the natural evolution of species in a multi-objective context. The evolution begins from a population of individuals generated at random, where each potential solution of the multi-objective optimization problem is represented by an individual. The fittest individuals are more likely to be selected to generate new ones by genetic operators. A Pareto ranking procedure and a crowding measure allow assessment of each individual in the current population. In the non-dominated sorting procedure, individuals are compared and ranked according to dominance, defining fronts that correspond to sets of non-dominated individuals. A rank is assigned to each front representing its level of domination. First, the best rank is given to the non-dominated individuals in the current population. Then, the second best rank is attributed the non-dominated individuals that are in the remaining population. This procedure is repeated until all individuals in the current population are included in a front. Then, the solutions of the worst-ranked non-dominated fronts are compared in terms of a crowding distance. Those in less crowded areas of the objective search space are selected for next generation.

The *gamultiobj* MATLAB function available in the Global Optimization Toolbox (MATLAB, 2019) will be used to approximate the Pareto optimal solutions. This function implements a variant of the elitist NSGA-II (Deb, 2001) in the multi-objective genetic algorithm. A set of algorithm options related to customizing the algorithm and termination criteria are provided. In this study, the default parameters of the MATLAB *gamultiobj* function were applied, where the size of the population was set to 50 individuals and the maximum number of generations of 400 was a stopping criterion. The vector of direction d is the same as in the single-objective problem. In all graphs presented, the sets of dominated and non-dominated solutions obtained among the 30 executions are drawn with points and circles, respectively. The representative solutions are marked as *.

9.6.3 RESULTS

The Pareto front for $SA\,vs.\,BT$ is depicted in Figure 9.14. It can be seen that from solution A to C, the SA value increases considerably, in contrast to the BT value, which decreases. From solution F to solution B, it is possible to observe a large decrease in BT 's value, maintaining this value for solution C. Solution L is dominated by solution C, since the values of SA and BT are lower.

The representative solutions (marked as black *) selected from the Pareto front are presented in Table 9.2.

Non-dominated solutions A to C presented in this pair of measures can be seen in Figure 9.13, and the orientations represented by F and L can be observed in Figure 9.15.

The Pareto front for $SA\,vs.\,RA$ is represented in Figure 9.16.

Table 9.3 presents some representative trade-offs from this Pareto curve. Solutions A and D are extremes of the Pareto front. Along the Pareto front curve, the value of

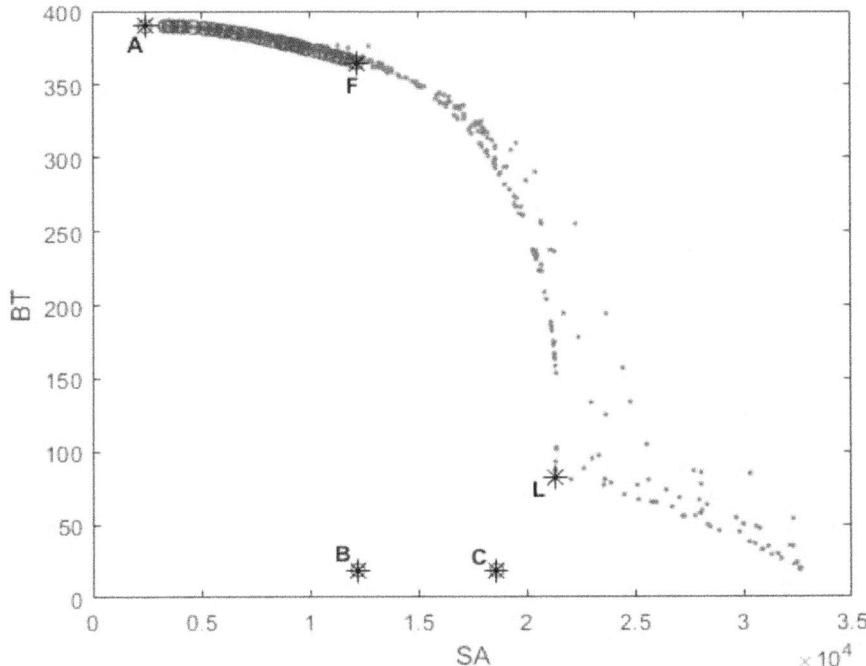

FIGURE 9.14   Pareto front of $SA\,vs.\,BT$.

TABLE 9.2
Representative Solutions for the Bi-Objective Problem for *SA vs. BT*

	θ_x	θ_y	SA	BT
A	$\forall \theta_x$	90.00	2.452647×10^3	3.895428×10^2
F	100.27	111.07	1.221222×10^4	3.649269×10^2
B	90.00	0.00	1.222754×10^4	1.851336×10^1

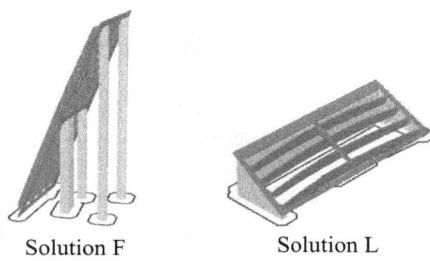

Solution F   Solution L

FIGURE 9.15   F and L representative solution orientations for car hoodvent model.

TABLE 9.3
Representative Solutions for the Bi-Objective Problem for *SA vs. RA*

	θ_x	θ_y	SA	RA
A	$\forall \theta_x$	90.00	2.452647×10^3	1.552228×10^1
G	147.96	90.06	3.260279×10^3	1.552131×10^1
H	125.90	77.97	7.444974×10^3	1.530164×10^1
D	134.31	115.70	1.290427×10^4	1.515983×10^1

SA increases and the value of *RA* decreases. It is also verified that solutions A to G show an increase in the value of *RA*, and for G to D, *RA* decreases. The orientations of the representative solutions A and D can be seen in Figure 9.13, and the solutions G and H are represented in Figure 9.17.

The graphical representation of the Pareto front for the combination of the objective functions *BT vs. RA* is depicted in Figure 9.18. Representative solutions are shown in the Pareto front and in Table 9.4, with a significant increase in the value

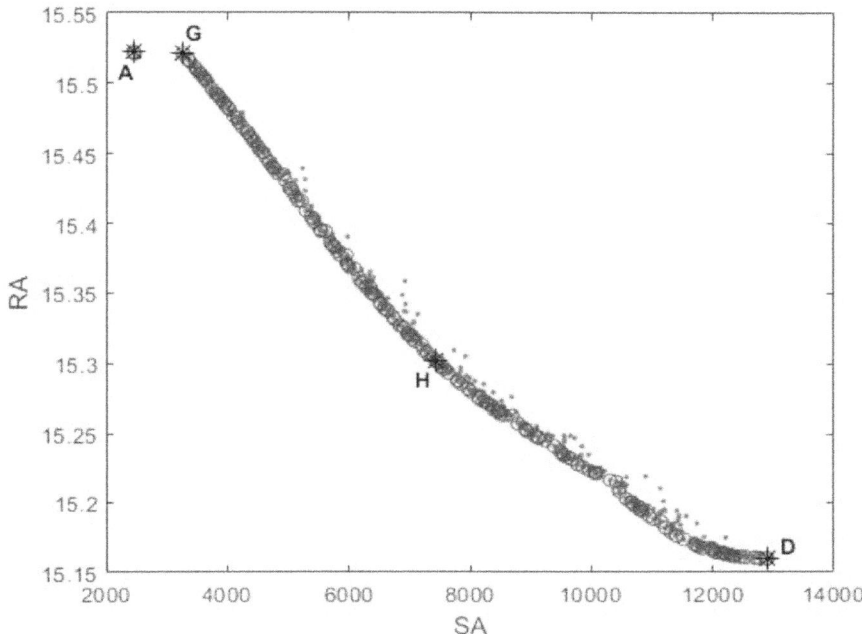

FIGURE 9.16 Pareto front of $SA\ vs.\ RA$.

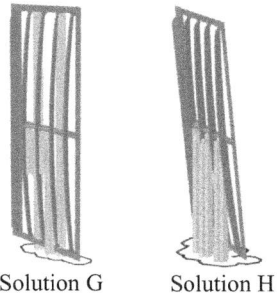

Solution G   Solution H

FIGURE 9.17 G and H representative solution orientations for car hoodvent model.

of BT from solution B to solution D. The BT function is inversely proportional to the RA function, as shown by the Pareto front curve: when the value of BT increases, the value of RA decreases. Representative solutions B, D and E are shown in Figure 9.13; the orientation presented by the letter L is represented in Figure 9.15; and solutions I, J and K can be seen in Figure 9.19.

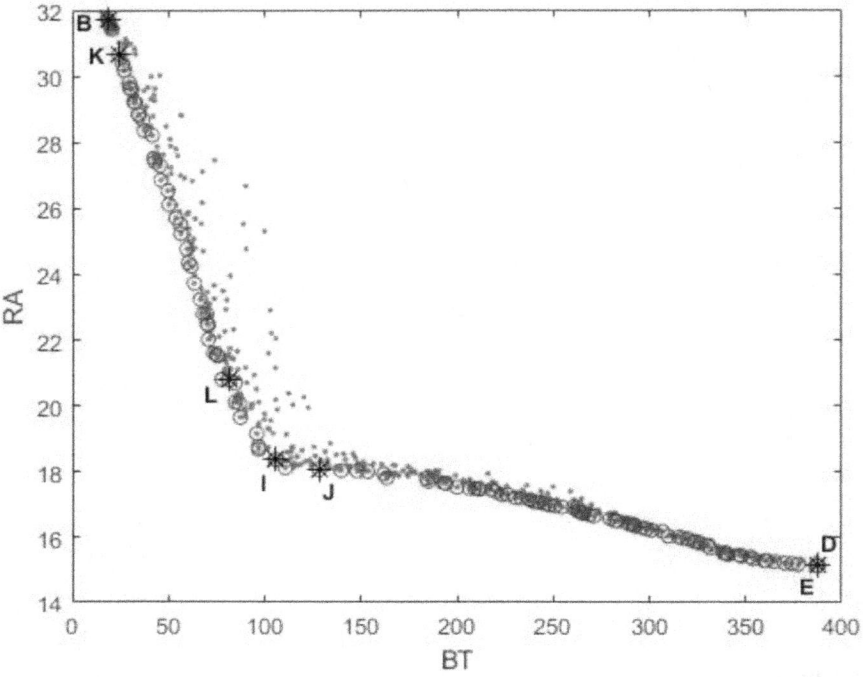

FIGURE 9.18   Pareto front of BT vs. RA.

TABLE 9.4

Representative Solutions for the Bi-Objective Problem for BT vs. RA

	θ_x	θ_y	BT	RA
B	90.00	180.00	1.851336×10^1	3.173028×10^1
K	92.56	0.04	2.438678×10^1	3.065574×10^1
L	117.87	0.77	8.120970×10^1	2.077520×10^1
I	130.85	0.78	1.054779×10^2	1.837226×10^1
J	134.59	176.56	1.282955×10^2	1.806146×10^1
E	134.31	64.29	3.878288×10^2	1.515999×10^1
D	134.31	115.70	3.878333×10^2	1.515983×10^1

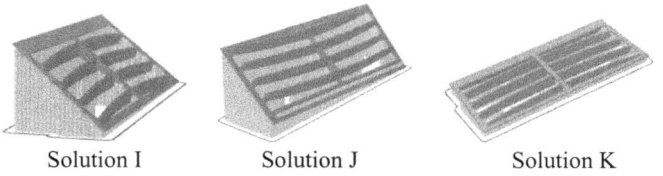

Solution I   Solution J   Solution K

FIGURE 9.19 I, J and K representative solution orientations for car hoodvent model.

9.7 MANY-OPTIMIZATION APPROACH

In this section, a many-objective approach, considering simultaneously optimizing the three objective functions ($SA\,vs.\,BT\,vs.\,RA$), is addressed.

The mathematical formulation of the many-objective optimization problem is expressed as

$$\min\left\{f_1\left(\theta_x,\theta_y\right),f_2\left(\theta_x,\theta_y\right),f_3\left(\theta_x,\theta_y\right)\right\}$$

$$s.t.\ \ 0\le\theta_x\le180 \tag{8}$$

$$a_k^*=\max\left\{a_k,et_{ij-1}^k\right\}$$

where f_1, f_2 and f_3 are the objective functions. Thus, in this problem (8), the three measures are simultaneously optimized, where $f_1 = SA$, $f_2 = BT$ and $f_3 = RA$. Like in the multi-objective approach, NSGA-II was also used to solve this problem, with the MATLAB function *gamultiobj*. In addition, stopping criteria were also the default values as well as the direction vector. Representative solutions A, B, C, D and E found in the single-objective and bi-objective problems were also found in this problem, as can be seen in the Pareto front represented in Figure 9.20.

In the following, the 2D projections for each pair of combinations of measures, SA and BT, SA and RA, and BT and RA, are shown in Figure 9.21.

Table 9.5 shows the values for the various objective functions. These different solutions are represented in Figure 9.13, Figure 9.15 and Figure 9.19.

9.8 DISCUSSION OF THE RESULTS

In all functions studied in this chapter, it can be seen that the minimizers of the three objective functions are different. Thus, these objectives are in conflict, and the several different trade-off solutions obtained during optimization represent different compromises between the objectives. In the three problems studied, several Pareto-optimal solutions were obtained, with 12 solutions chosen as representative solutions. Solutions A, B, C, D and E were found in all problems (single-objective, bi-objective and many-objective problems) and solutions J, K

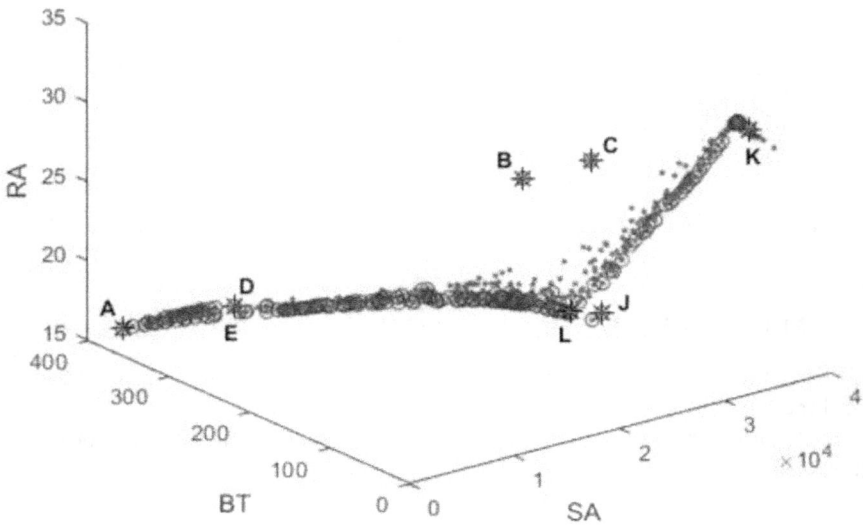

FIGURE 9.20   Pareto front of $SA\,vs.\,BT\,vs.\,RA$.

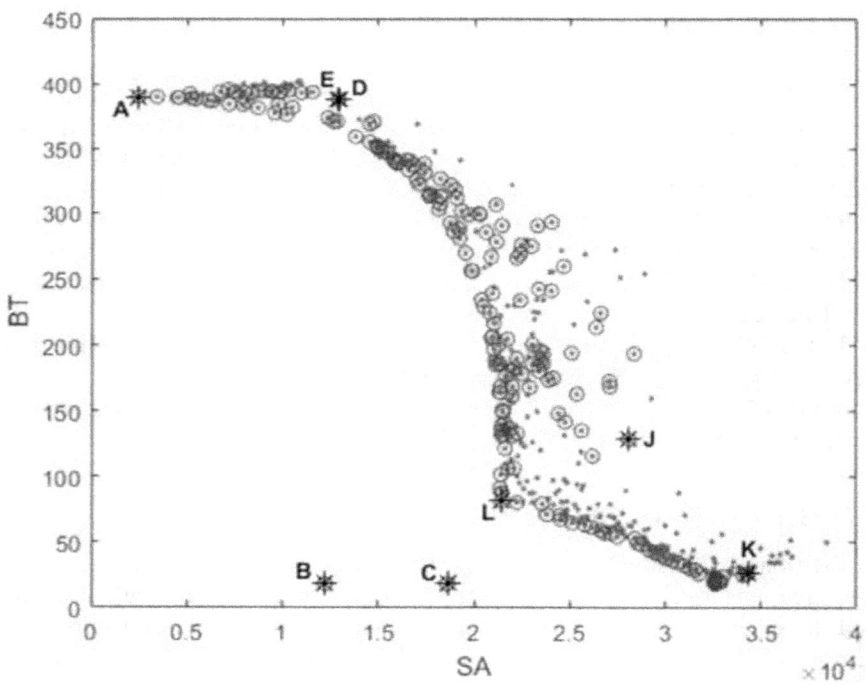

2D projection of SA and BT

FIGURE 9.21   2D projected solutions of the Pareto front of $SA\,vs.\,BT\,vs.\,RA$.

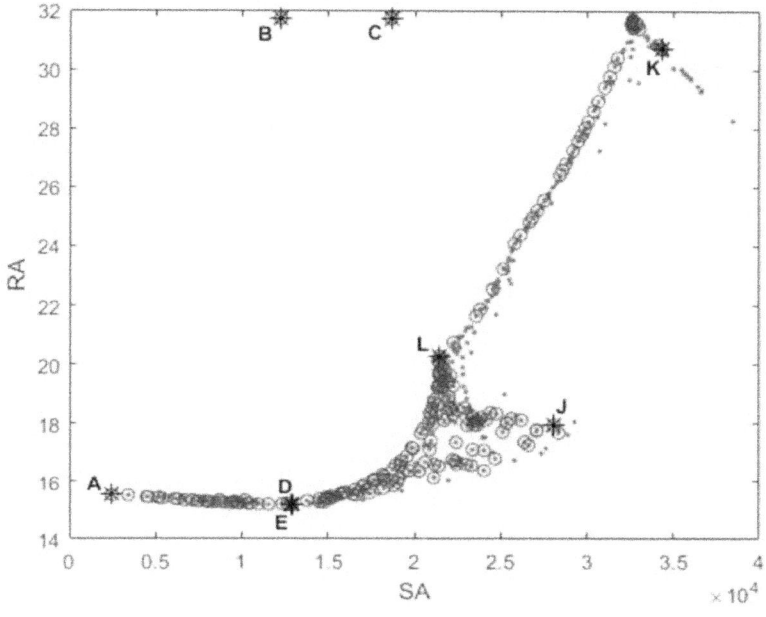

2D projection of *SA* and *RA*

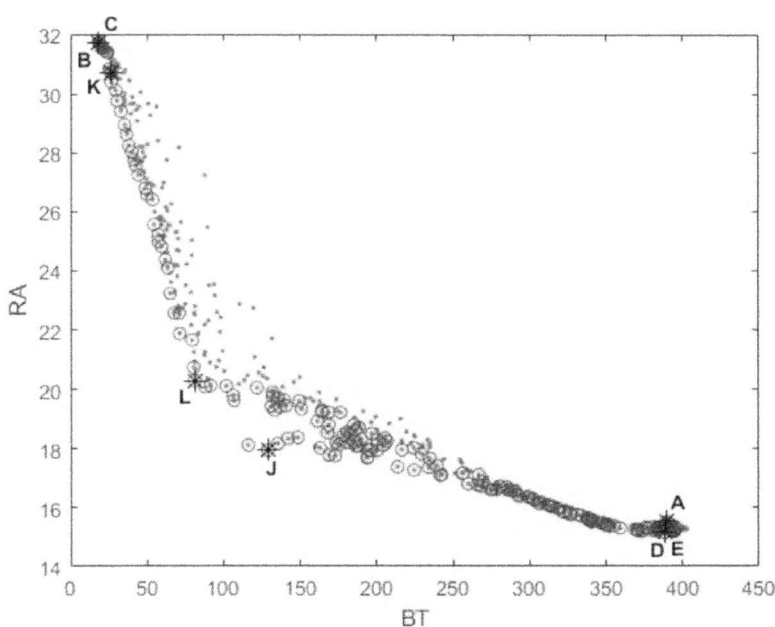

2D projection of *BT* and *RA*

FIGURE 9.21   (*Continued*)

TABLE 9.5

Representative Solutions for the Many-Objective Problem for SA vs. BT vs. RA

	θ_x	θ_y	SA	BT	RA
A	$\forall \theta_x$	90.00	2.452647×10^3	3.895428×10^2	1.552228×10^1
B	90.00	180.00	1.222754×10^4	1.851336×10^1	3.173028×10^1
E	134.31	64.29	1.290403×10^4	3.878288×10^2	1.515999×10^1
D	134.31	115.70	1.290427×10^4	3.878333×10^2	1.515983×10^1
C	90.00	0.00	1.861518×10^4	1.851335×10^1	1.909200×10^1
L	117.87	0.77	2.132272×10^4	8.120970×10^1	2.077520×10^1
J	134.59	176.56	2.636640×10^4	1.282955×10^2	1.806146×10^1
K	92.56	0.04	3.192372×10^4	2.438678×10^1	3.065574×10^1

and L were found only in bi-objective and many-objective problems. In addition to these, other solutions were selected that were found in the bi-objective problem. Solution F was found in the SA vs. BT problem, G and H were found in the combination of objective functions SA vs. RA and solution I was found in the BT vs. RA problem. Solution L was found in the bi-objective problem SA vs. BT as a dominated solution. In addition to this combination of functions, solution L was also obtained in the bi-objective problem BT vs. RA and in the many-objective problem SA vs. BT vs. RA as a non-dominated solution. Solution A has the same function values ($\forall \theta_x$, 90.00).

Through the values of the objective functions, the path graph presented in Figure 9.22 and the orientations allow identification of trade-offs and comparison of the characteristics of the representative solutions. It is possible to observe that the orientation obtained in solution A is the one with the least support and has the least roughness, with the longest build time. Conversely, the values of SA and RA are maximum for solution K; that is, it has more support, and it is an orientation that increases the surface roughness of the model. Solutions B and C are the orientations that spend less time on printing; that is, they have a shorter height (they lie on the 3D printing platform horizontally). The longest build time is in solution H, which implies a maximum printing height.

9.9 CONCLUSIONS AND FUTURE WORK

In this chapter, the build orientation optimization problem of the car hoodvent model in additive manufacturing is studied. The optimization of the build orientation aims to reduce the construction costs and improve the quality of the car hoodvent model. First, a single-objective approach, based on GA, was implemented to separately solve three objective functions: the support area, the build time and the surface roughness.

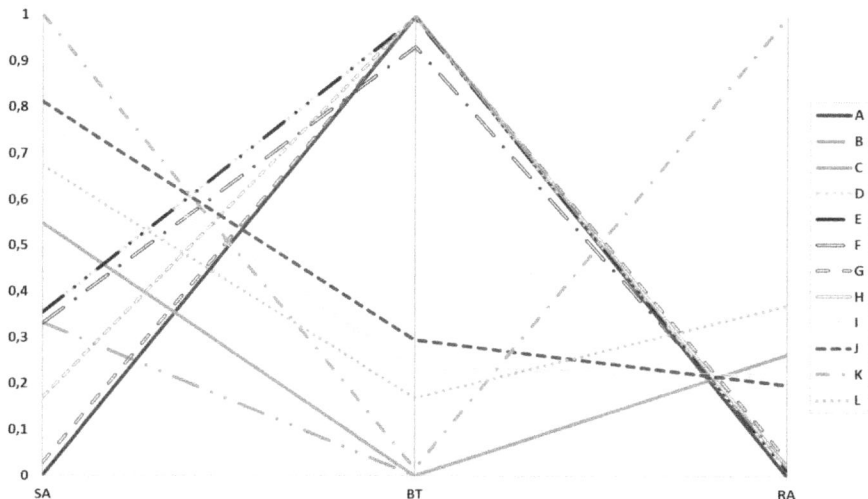

FIGURE 9.22   Path graph for the three objective functions for the car hoodvent model.

Then, a bi-objective approach was used to solve each combination of two objective functions. Finally, a many-objective approach simultaneously solves the three objective functions. The NSGA-II algorithm, provided in the MATLAB environment, was applied for the bi-objective and many-objective problems. The objective of this chapter is to compute the optimal build orientation of a 3D model in order to assist the decision-maker in choosing the orientation considering different measures or factors.

Initially, a single-objective problem was presented using the three objectives, *SA*, *BT* and *RA*, and the orientations A, B, C, D and E were found. Then, a bi-objective approach with combinations of two measures was studied, *SA vs. BT*, *SA vs. RA* and *BT vs. RA*. The Pareto fronts obtained for each combination of the three objective functions were shown, and the representative non-dominated solutions were selected. In the third part of this chapter, a multi-objective problem with the objective of optimizing the three quality measures simultaneously, *SA vs. BT vs. RA*, was proposed. In this phase, the 3D Pareto front of the three objective functions and the 2D projections of this were shown in order to select the representative solutions. A path value graph representing the objectives and the representative solutions was used to analyze the obtained solutions. This graph allows the identification of the compromises between the objective functions. This approach permitted us to obtain different optimal solutions successfully, taking into account the criteria based on the three quality measures. This strategy proved promising for the future, since it helps the decision-maker select a build orientation according to his/her preferences.

After analyzing the results of all the studied problems, and since the Pareto fronts presented non-convexities and discontinuities, the importance of formulating and solving these kinds of problems as multi-objective ones was highlighted. In addition, this approach allowed us to perceive the significance and relevance for the industry to print a 3D object in a short time, at a reduced cost, wasting less material and improving the surface finish. In the future, other difficult 3D models will be tested, performing multi-objective optimization and additional quality measures. Furthermore, multi-criteria decision-making methods will be investigated in order to find explicit solution orders and preferences.

ACKNOWLEDGMENTS

This work has been developed under the FIBR3D project Hybrid processes based on additive manufacturing of composites with long or short fibers reinforced thermoplastic matrix (POCI-01–0145-FEDER-016414), supported by the Lisbon Regional Operational Programme 2020, under the PORTUGAL 2020 Partnership Agreement, through the European Regional Development Fund (ERDF). This work has been supported by FCT—Fundação para a Ciência e Tecnologia within the R&D Units Project Scope: UIDB/00319/2020 and UIDB/05757/2020.

NOTES

1 Available at: https://commons.wikimedia.org/wiki/File:Rapid_prototyping_ worldwide_ by_Zureks.png, Accessed: 2020–12–28.
2 Available at: https://formlabs.com/blog/additive-manufacturing-vs-subtractive-manufacturing/, Accessed: 2020–12–28.

REFERENCES

Abdulhameed, O., Al-Ahmari, A., Ameen, W., and Mian, S. H. (2019). Additive manufacturing: Challenges, trends, and applications. *Advances in Mechanical Engineering*, 11(2):1687814018822880.

Ahn, D.-K., Kwon, S.-M., and Lee, S.-H. (2008). Expression for surface roughness distribution of FDM processed parts. In *2008 International Conference on Smart Manufacturing Application,* pages 490–493. IEEE.

Alexander, P., Allen, S., and Dutta, D. (1998). Part orientation and build cost determination in layered manufacturing. *Computer-Aided Design*, 30(5):343–356.

Behnam, N. (2011). *Surface Roughness Estimation for FDM Systems*. PhD thesis, MSc thesis, Ryerson University, Toronto.

Bikas, H., Stavropoulos, P., and Chryssolouris, G. (2016). Additive manufacturing methods and modelling approaches: A critical review. *The International Journal of Advanced Manufacturing Technology*, 83(1–4):389–405.

Brika, S. E., Zhao, Y. F., Brochu, M., and Mezzetta, J. (2017). Multi-objective build orientation optimization for powder bed fusion by laser. *Journal of Manufacturing Science and Engineering*, 139(11):111011.

Byun, H. S., and Lee, K. H. (2006). Determination of optimal build direction in rapid prototyping with variable slicing. *The International Journal of Advanced Manufacturing Technology*, 28(3–4):307.

Campbell, R. I., Martorelli, M., and Lee, H. S. (2002). Surface roughness visualisation for rapid prototyping models. *Computer-Aided Design*, 34(10):717–725.

Canellidis, V., Dedoussis, V., Mantzouratos, N., and Sofianopoulou, S. (2006). Pre-processing methodology for optimizing stereolithography apparatus build performance. *Computers in Industry*, 57(5):424–436.

Canellidis, V., Giannatsis, J., and Dedoussis, V. (2009). Genetic-algorithm-based multi-objective optimization of the build orientation in stereolithography. *The International Journal of Advanced Manufacturing Technology*, 45(7–8):714–730.

Cerneels, J., Voet, A., Ivens, J., and Kruth, J.-P. (2013). Additive manufacturing of thermoplastic composites. *Composites Week@ Leuven*, pages 1–7.

Das, P., Mhapsekar, K., Chowdhury, S., Samant, R., and Anand, S. (2017). Selection of build orientation for optimal support structures and minimum part errors in additive manufacturing. *Computer-Aided Design and Applications*, 14(sup1):1–13.

Deb, K. (2001). *Multi-Objective Optimization Using Evolutionary Algorithms*. John Wiley & Sons, Inc., New York.

Del Re, F., Scherillo, F., Contaldi, V., Palumbo, B., Squillace, A., Corrado, P., and Di Petta, P. (2019). Mechanical properties characterisation of alsi10mg parts produced by laser powder bed fusion additive manufacturing. *International Journal of Materials Research*, 110(5):436–446.

Di Angelo, L., Di Stefano, P., Dolatnezhadsomarin, A., Guardiani, E., and Khorram, E. (2020). A reliable build orientation optimization method in additive manufacturing: The application to FDM technology. *International Journal of Advanced Manufacturing Technology*, 108.

Ford, S., and Despeisse, M. (2016). Additive manufacturing and sustainability: An exploratory study of the advantages and challenges. *Journal of Cleaner Production*, 137: 1573–1587.

Gok, A. (2015). A new approach to minimization of the surface roughness and cutting force via fuzzy topsis, multi-objective grey design and RSA. *Measurement*, 70:100–109.

Goldberg, D.E. (1989). *Genetic Algorithms in Search Optimization, and Machine Learning*. Addison Wesley Publishing Co. Inc., New York.

Gurrala, P.K., and Regalla, S.P. (2014). Multi-objective optimisation of strength and volumetric shrinkage of FDM parts: A multi-objective optimization scheme is used to optimize the strength and volumetric shrinkage of FDM parts considering different process parameters. *Virtual and Physical Prototyping*, 9(2):127–138.

Jaiswal, P., Patel, J., and Rai, R. (2018). Build orientation optimization for additive manufacturing of functionally graded material objects. *The International Journal of Advanced Manufacturing Technology*, 1–13.

Jiang, J., Xu, X., and Stringer, J. (2018). Support structures for additive manufacturing: A review. *Journal of Manufacturing and Materials Processing*, 2(4):64.

Jibin, Z. (2005). Determination of optimal build orientation based on satisfactory degree theory for rpt. In *Computer Aided Design and Computer Graphics, 2005. Ninth International Conference on Computer Aided Design and Computer Graphics (CAD-CG'05)*, pages 6—pp. IEEE. doi: 10.1109/IEEECONF11203.2005

Khodaygan, S., and Golmohammadi, A. H. (2018). Multi-criteria optimization of the part build orientation (PBO) through a combined meta-modeling/NSGAII/TOPSIS method for additive manufacturing processes. *International Journal on Interactive Design and Manufacturing (IJIDeM)*, 12(3):1071–1085.

Kruth, J. P. (1991). Material increases manufacturing by rapid prototyping techniques. *CIRP Annals-Manufacturing Technology*, 40(2):603–614.

Lee, J. Y., An, J., and Chua, C. K. (2017). Fundamentals and applications of 3D printing for novel materials. *Applied Materials Today*, 7:120–133.

Li, A., Zhang, Z., Wang, D., and Yang, J. (2010). Optimization method to fabrication orientation of parts in fused deposition modeling rapid prototyping. In *2010 International Conference on Mechanic Automation and Control Engineering*, pages 416–419. IEEE. doi: 10.1109/MACE16739.2010

Mason, A. (2007). *Multi-Axis Hybrid Rapid Prototyping Using Fusion Deposition Modeling*. ProQuest. Doctoral dissertation.

Masood, S., Rattanawong, W., and Iovenitti, P. (2003). A generic algorithm for a best part orientation system for complex parts in rapid prototyping. *Journal of Materials Processing Technology*, 139(1–3):110–116.

MATLAB (2019). *Version 9.6.0.1214997 (R2019a)*. The MathWorks Inc., Natick, MA.

Matos, M. A., Rocha, A. M. A. C., Costa, L. A., and Pereira, A. I. (2019a). A multi-objective approach to solve the build orientation problem in additive manufacturing. In *International Conference on Computational Science and Its Applications*, pages 261–276. Springer, Berlin.

Matos, M. A., Rocha, A. M. A. C., and Pereira, A. I. (2019b). On optimizing the build orientation problem using genetic algorithm. In *AIP Conference Proceedings*, volume 2116, page 220006. AIP Publishing LLC, Melville.

Matos, M. A., Rocha, A. M. A. C., and Pereira, A. I. (2020). Improving additive manufacturing performance by build orientation optimization. *The International Journal of Advanced Manufacturing Technology*, 1–13.

Matos, M. A., Rocha, A. M. A. C., and Costa, L. A. (2021a). Many-objective optimization of build part orientation in additive manufacturing. *The International Journal of Advanced Manufacturing Technology*, 112(3), 747–762.

Matos, M. A., Rocha, A. M. A. C., Costa, L. A., and Pereira, A. I. (2021b). Multi-objective optimization in the build orientation of a 3D CAD model. In *Advances in Evolutionary and Deterministic Methods for Design, Optimization and Control in Engineering and Sciences*, pages 99–114. Springer, Berlin.

Mele, M., and Campana, G. (2020). Sustainability-driven multi-objective evolutionary orienting in additive manufacturing. *Sustainable Production and Consumption*, 23, 138–147.

Mohan Pandey, P., Venkata Reddy, N., and Dhande, S. G. (2003). Slicing procedures in layered manufacturing: A review. *Rapid Prototyping Journal*, 9(5):274–288.

Nakano, A. L., and Cunha, I. L. L. (2007). *Transformaçoes geométricas 2D e 3D*. pages 1–25.

Ngo, T. D., Kashani, A., Imbalzano, G., Nguyen, K. T., and Hui, D. (2018). Additive manufacturing (3Dprinting): A review of materials, methods, applications and challenges. *Composites Part B: Engineering*, 143:172–196.

Olsen, J., and Kim, I. Y. (2020). Design for additive manufacturing: 3D simultaneous topology and build orientation optimization. *Structural and Multidisciplinary Optimization*, 62:1989–2009.

Padhye, N., and Deb, K. (2011). Multi-objective optimisation and multi-criteria decision making in SLS using evolutionary approaches. *Rapid Prototyping Journal*, 17(6):458–478.

Pandey, P., Reddy, N. V., and Dhande, S. (2007). Part deposition orientation studies in layered manufacturing. *Journal of Materials Processing Technology*, 185(1–3):125–131.

Pandey, P. M. (2010). Rapid prototyping technologies, applications and part deposition planning. *Retrieved October*, 15:550–555.

Pandey, P. M., Reddy, N. V., and Dhande, S. G. (2003). Improvement of surface finish by staircase machining in fused deposition modeling. *Journal of Materials Processing Technology*, 132(1–3):323–331.

Pereira, S., Vaz, A. I. F., and Vicente, L. N. (2018). On the optimal object orientation in additive manufacturing. *The International Journal of Advanced Manufacturing Technology*, 98(5):1685–1694.

Phatak, A. M., and Pande, S. (2012). Optimum part orientation in rapid prototyping using genetic algorithm. *Journal of Manufacturing Systems*, 31(4):395–402.

Qin, Y., Qi, Q., Scott, P. J., and Jiang, X. (2019). Determination of optimal build orientation for additive manufacturing using Muirhead mean and prioritised average operators. *Journal of Intelligent Manufacturing*, 30(8):3015–3034.

Raju, B., Shekar, U. C., Venkateswarlu, K., and Drakashayani, D. (2014). Establishment of process model for rapid prototyping technique (stereolithography) to enhance the part quality by Taguchi method. *Procedia Technology*, 14:380–389.

Rocha, A. M. A. C., Pereira, A. I., and Vaz, A. I. F. (2018). Build orientation optimization problem in additive manufacturing. In *International Conference on Computational Science and Its Applications*, pages 669–682. Springer, Berlin.

Shanmugam, V., Das, O., Babu, K., Marimuthu, U., Veerasimman, A., Johnson, D. J., Neisiany, R. E., Hedenqvist, M. S., Ramakrishna, S., and Berto, F. (2021). Fatigue behaviour of FDM-3D printed polymers, polymeric composites and architected cellular materials. *International Journal of Fatigue*, 143:106007.

Shen, H., Ye, X., Xu, G., Zhang, L., Qian, J., and Fu, J. (2020). 3D printing build orientation optimization for flexible support platform. *Rapid Prototyping Journal*, 26(1), 59–72, Emerald.

Sheoran, A. J., and Kumar, H. (2020). Fused Deposition modeling process parameters optimization and effect on mechanical properties and part quality: Review and reflection on present research. *Materials Today: Proceedings*, 21:1659–1672.

Taufik, M., and Jain, P. K. (2013). Role of build orientation in layered manufacturing: A review. *International Journal of Manufacturing Technology and Management*, 27(1–3):47–73.

Thrimurthulu, K., Pandey, P. M., and Reddy, N. V. (2004). Optimum part deposition orientation in fused deposition modeling. *International Journal of Machine Tools and Manufacture*, 44(6):585–594.

Wang, W., Chao, H., Tong, J., Yang, Z., Tong, X., Li, H., . . . and Liu, L. (2015). Saliency-preserving slicing optimization for effective 3D printing. *Computer Graphics Forum*, 34(6):148–160.

Zhang, Y., Bernard, A., Harik, R., and Karunakaran, K. (2017). Build orientation optimization for multi-part production in additive manufacturing. *Journal of Intelligent Manufacturing*, 28(6):1393–1407.

Zhang, Y., De Backer, W., Harik, R., and Bernard, A. (2016). Build orientation determination for multi-material deposition additive manufacturing with continuous fibers. *Procedia CIRP*, 50(2016):414–419.

10 Test Bed Problem
Case Study—Design for Customization of an Assistive Device for Daily Activities

Zilda de Castro Silveira

CONTENTS

10.1 DEMAND CONTEXT

An assistive device can be described as any item, piece of equipment or product system that can be purchased commercially, modified or customized and is used to increase, maintain or improve the functional capacity of individuals with disabilities.

DOI: 10.1201/9781003123866-10

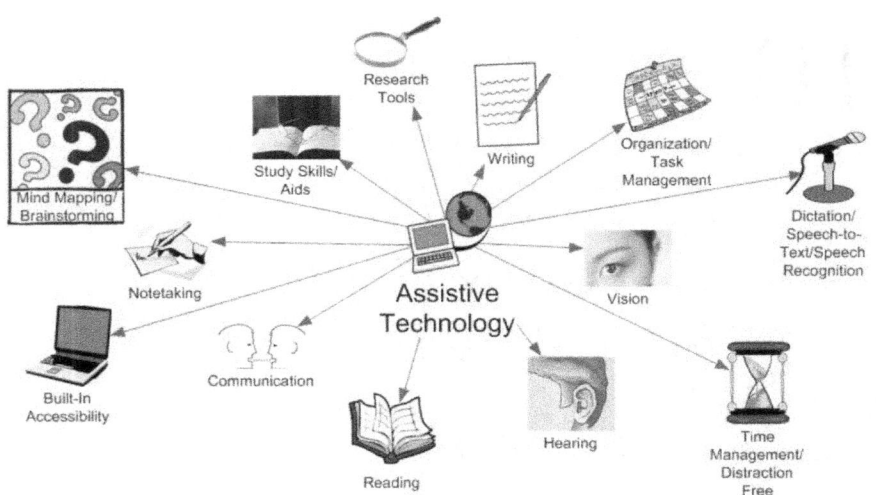

FIGURE 10.1   Categories of technical aids in assistive technology.

Source: Free or Low Cost Assistive Technology for Everyone—CLASS Disability Resources I Augsburg University. Access: 5/17/2021

Squires et al. (2019) stated that, in addition to greater independence, there is positive evidence of physical, psychological and economic impacts of assisted technology (AT) on quality of life, social inclusion and reduced costs of care. Technological assisted resources can also be extended to education and formal employment, generating multiple and life-changing benefits. (Stumbo et al., 2009). Assistive technology can be divided into 12 categories of technical aids. Figure 10.1 presents a visual representation (from Augsburg University, https://www.augsburg.edu/) of these kinds of technical aids. Harris and Smith (2017, p. 24) stated that in the past ten years, "user-centered design" has become an essential component of any research proposal that seeks to develop enabling technology for people living with challenging healthcare (assistive technology). Gherardini et al. (2018) affirmed that co-design proposals offer a more adequate answer to the functional and psychological needs of each individual because he or she is more directly involved in the process, thus contributing to the reduction of the stigmatization of the product by the user and, therefore, consequently reducing the abandonment rates of ADs.

De Couvreur and Goosens (2011) proposed the design for (every) one model, based on the concept of co-design and characterized by horizontal innovation networks, which mix principles of universal design and therapeutic/rehabilitation activities. Although the authors have made significant and innovative contributions, mainly with regard to composing a base community for the design integrated with the local context, there is a gap in the understanding of the application of strategies to generate solutions.

Coton et al. (2014) suggested an iterative and multidisciplinary methodology, which considers human and environmental factors in design development. The definition of a multi-dimensional competence and a balance between design requirements and technical solutions from users and technology domains are aspects of the paper that stand out. However, the description of the specific design, as well the use of design tools, presents a gap. Following similar principles, Chavarriaga et al. (2014) described a multidisciplinary approach classified by three steps: 1) characterization of the problem considering socio-cultural aspects of the target audience, 2) systematic review and technological assistive device solutions and 3) community building through the interaction of stakeholders. The authors mention the use of design techniques such as the theory of inventive problem solving, the Russian acronym for TRIZ (Theory of Inventive Problem Solving) and the axiomatic design approach (Chavarriaga et al., 2014), but the absence of physical prototypes to validate the generated concepts ends up limiting the user's participation (Stappers et al., 2011).

Ostuzzi et al. (2015) was concerned with detailing the different actors involved and their corresponding roles for each of the stages of the implemented methodology, characterized by iterative cycles of ideation, prototyping and testing. Gherardini et al. (2018) also contributed an interesting co-design approach, with emphasis on the use of tools for assessing familiarity and satisfaction with the device, such as the Psychosocial Impact of Scale Assistive Devices (PIADS) and Québec User Evaluation of Satisfaction with Technical Aids (B-QUEST). In the context of investigating abandonment and the relationship with design structure for assistive technology, Maia and Freitas (2014) identified that designs are developed in an unsystematic way, disregarding, for example, aesthetic parameters of the AD for the user, as well as a lack of user participation in fundamental decision-making design steps and lack of application of tools for AD assessment. In conclusion, the authors identified large lacks in important stages of the design process, especially in generating ideas, testing and training.

Thus, most of literature related to ADs presents studies with a focus on the technological domain with some application of health assessment tools but without iteration and interactions of the actors between the design phases. Design solutions cited in these papers included 3D scanning of the anatomical region, geometric modeling, some numerical simulations using CAD/CAE systems and in some cases the use of low-cost 3D printing equipment to manufacture the functional prototype (Portnova et al., 2018). Despite the growing demand for AT products, in view of the worldwide trend of population aging, many challenges are still identified in this area, such as the lack of correspondence between demand and supply; users' needs, which are often little known and unsatisfactorily attended to; problems with product reliability; and the stigmatization caused by poor design (Ulrich et al., 2008). The lack of specialized distribution networks, multiple profit margins and, in many cases, the monopolistic market are other difficulties to be mentioned (Plos et al., 2012).

This lack of AD is emphasized by De Couvreur and Goosens (2011), who related that approximately 600 million individuals with disabilities living with assistive devices (ADs) are not adapted or do not have access to them. Mihailidis and Polgar (2016) pointed out the development of AD designs in an unsystematic way, with many adaptations of already consolidated solutions, leads to technological stagnation in assistive technology products. Aligned with other factors, such as cultural and economic factors, the lack of systematization in the development of assistive products means there is a high rate of abandonment of ADs. Phillips and Zhao (1993) identified a 29.3% dropout rate for devices of various types, associated with four main aspects: (1) disregard for the user's opinion, (2) difficulties in acquiring ADs, (3) poor device performance and (4) changing user needs or priorities. Because of this abandonment, one can mention the gradual residual loss of the user's functional abilities. With regard to the economic aspect, there is an increase in expenses due to ineffective use of government resources and other providers, resulting in a high number of AD disposals in many countries.

On the other hand, with the gradual increase in the demand for ADs together with better adaptations to the user, a very favorable niche is created for collaboration between professionals in the fields of engineering and health, specifically occupational therapists (Maia & Freitas, 2014). In this case, engineers contribute with their knowledge of technology and product development methods, while healthcare professionals contribute with their user-centered practices and tools for evaluating the quality and performance of assistive technology design attributes.

10.2 DEVELOPMENT OF A CUSTOMIZED ASSISTIVE DEVICE FOR A SELF-SUPPORTING UTENSIL FOR PEOPLE WITH PARKINSON'S DISEASE AND ESSENTIAL TREMOR

The development of an AD product presents the same nature as a customized design. The use of a design methodology can incorporate users in design decision-making, including healthcare professionals, in the initial design phases related to informational and conceptual design. User participation in the interaction and iteration processes in the testing and refinement phases of the conceptual solution can contribute significantly to generating products with a lower probability of abandonment and also fulfill most of the attributes foreseen for AT: safety, comfort, independence and ergonomics. In this context, in this chapter, we will present a case study where design methodology was based on the AT-8esign (AT-8d) proposed by Santos and Silveira (2020), which was partially supported by traditional design methodologies from Ullman (2010). The AT-8esign methodology has iterative and dynamic approaches and aims to apply engineering techniques in assistive device design, with a user-centered approach and participatory design. The steps and tasks are presented in Figure 10.2(a) and (b).

The first design step, called design cross-domain (DCD)—creates a favorable interdisciplinary domain, which is highlighted by the integration between the various stakeholders involved, with their personal experiences and needs, technical knowledge and skills, languages and communication strategies. As a result, a suitable design structure

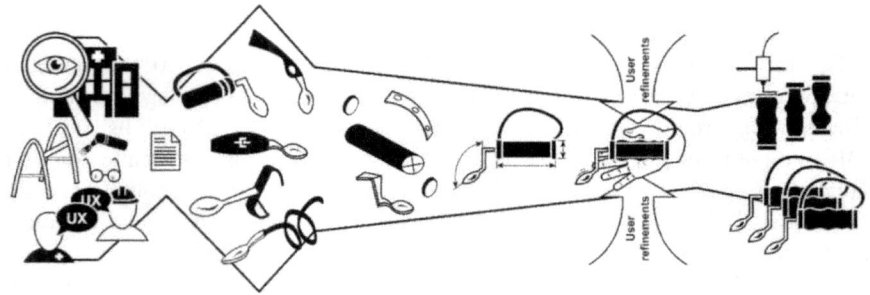

FIGURE 10.2   AT-8esign design methodology.

Source: Santos & Silveira, 2020

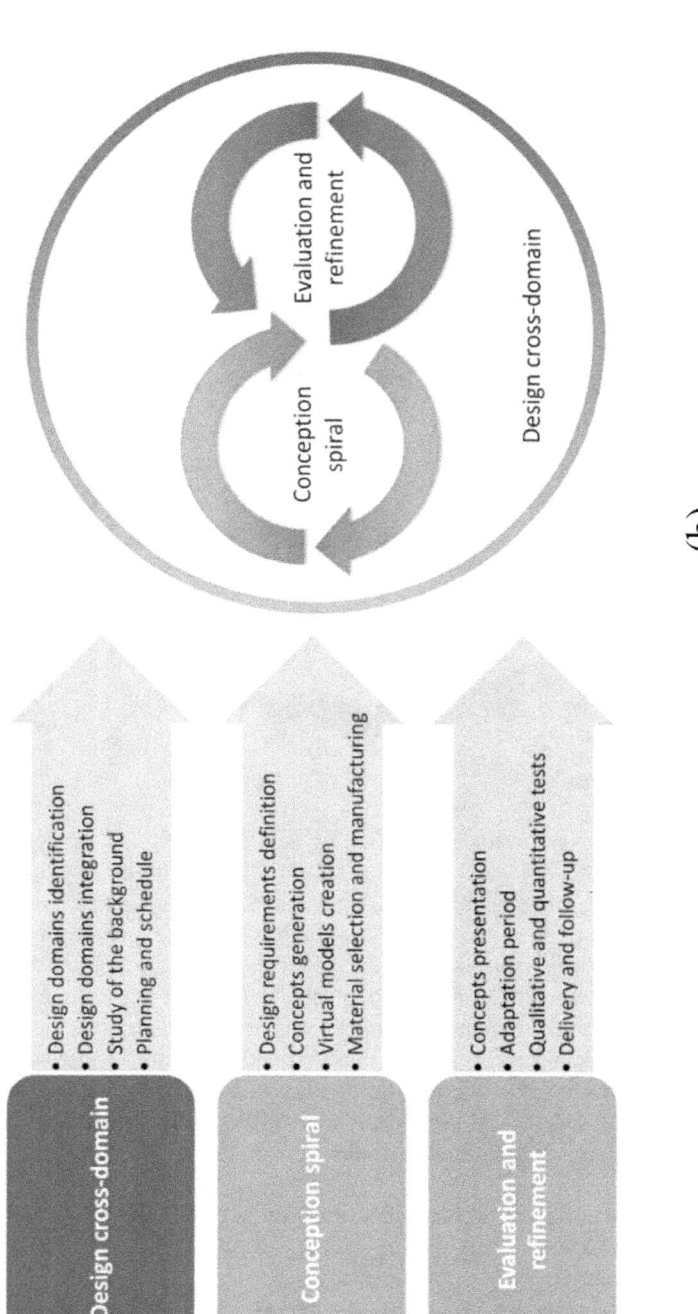

FIGURE 10.2 (*Continued*)

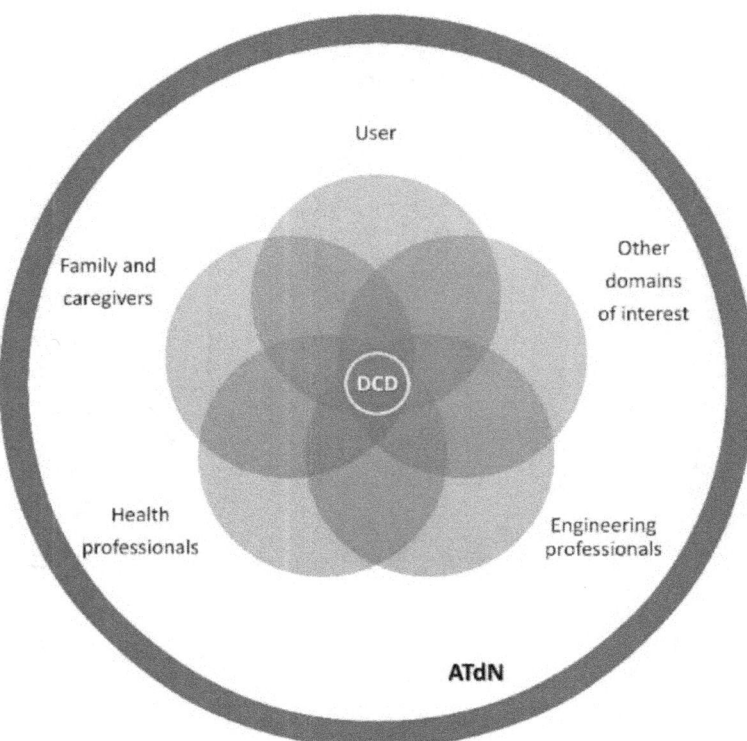

FIGURE 10.2 *(Continued)*

is created for the conception and generation of new AT ideas and conceptual solutions and identifying technological gaps. The main DCD step [Figure 10.2(c)] is to give design visions, values, language and goals composing an assistive technology design nucleus (ATdN) domain. The second phase—the design spiral—is the most creative core of the process, in which the concepts and solutions for AD are elaborated upon and improved, with the support of design methods, and in which the concepts and solutions are manufactured via additive manufacturing (AM), virtual and physical mock-ups and functional and technical prototypes and products. Finally, the third phase—evaluation and refinement—aims to assess with the user and other members of the DCD the satisfaction with the DA, as well as its functionality and effectiveness, in order to refine the previous phase in search of design improvement.

1. Linearized evolution of problem identification to the final product (adapted from Ullman, 2010).
2. AT-8esign structure.
3. Interactive phase of a design for customization so-called design cross-domain.

With a brief explanation of the design methodology used, the case study will be presented.

10.2.1 CROSS-DOMAIN DESIGN

10.2.1.1 Design Domain Identification

This study was started by a functional demand identified by research in occupational therapy at the Rehabilitation Center at Hospital de Clinicas—Federal University of Triângulo Mineiro, Brazil. Pachelli (2014) performed a comparative study of occupational performance between a conventional spoon and an adapted spoon, both designed for eating using the experimental design AB single case, which is a two-part or phase design composed of a baseline ("A" phase) with no changes and a treatment or intervention ("B") phase. The sample consisted of an individual man, 60 years old, with Parkinson's disease (PD). The Canadian Occupational Performance Measure (COPM) questionnaire was applied to evaluate satisfaction level by number of spoonfuls in eating activities with the conventional spoon model and with the adapted spoon. The results showed performance measures: the number of spoonfuls significantly increased with the adapted spoon, and satisfaction with the performance was not statistically significant. Pachelli (2014) concluded that the adapted spoon model significantly improved occupational performance of the research participant.

10.2.1.2 Design Domain Integration

PD is classified as the second most prevalent neurodegenerative disease. Parkinson's disease is defined as a degenerative health condition of the central nervous system related to reduction of dopamine production. Recent studies have associated PD with increased age, and it currently affects approximately 1% of the population over 60 years of age (Tysenes & Storstein, 2017). PD is a major challenge for public health, either due to the aging of the population or due to the different degrees of disability, stigmatization and restriction of social participation. Additionally, 700,000 new cases of PD are expected to be diagnosed by 2040, based on data related to aging alone (Rossi et al., 2018; Majeed et al., 2021)

Regarding the functional aspects and the restrictions to involvement in significant activities, the person with PD presents limitations in different activities of daily living (ADLs), having difficulties specifically in writing, dressing themselves and eating performance and satisfaction. Considering eating specifically, the literature suggests that people with PD have difficulties handling utensils due to their motor deficits, tremors and bradykinesia (Ma et al., 2008; Ma et al., 2009; Sabari et al., 2019). Eating can be described as a fundamental activity for survival and basic well-being (Townsend & Polatajko, 2007). It represents an important capability for the maintenance of social relations, as it enables a person to interact with family and friends, be more involved in daily routines, establish affective bonds, strengthen habits, share cultural moments and increase social participation.

Many utensils have been developed and are available to deal with the different demands of people with PD, such as the weighted spoon, the built-up handle weighted spoon, the swivel spoon and the liftware steady spoon (Sabari et al., 2019). The main features of these devices are an increase in the handle diameter, weight variation, the angle of the tip of the spoon and a mechanical counterbalance. For people with PD, the limited rehabilitation literature suggests that weighted cutlery is the most appropriate for use (Forwell et al., 2008; Schultz-Krohn et al., 2005). However, despite the limited amount of research on the subject, some studies point to the influence of utensil weight on the eating task, recommending the use of light cutlery when the activity demands fast movement (Hewer et al., 1972; Homberg et al., 1987; Ma et al., 2009; Meshack & Norman, 2002). This kind of AD is restricted in the Brazilian market, with some adaptations made empirically. Importing this type of AD is difficult for low-income users due to the high price, and it often does not aid with impaired function. According to Centro de Gestão e Estudos Estratégicos-Brazil, in 2012, the availability of AD on the Brazilian market is still insufficient.

Pachelli's (2014) work provided an initial survey of commercial options for eating utensils and an indication from the literature of the potential use of counterweights for people affected by PD. Finally, a design demand was identified for an AD focused on support for self-feeding in order to fulfill the functions of comfort, ergonomics, safety and independence for the early stages of Parkinson's disease and have a low cost.

10.2.1.3   Study Background

In order to complement scientific papers and commercial products intended for people with PD, a survey was conducted of a patent database as well as an updated search for commercial adapted utensils for self-feeding.

First, physical and virtual stores that operate in the assistive products segment were listed, and keywords in Portuguese and English were used to survey commercial solutions. Table 10.1 presents a summary of the main adapted utensils identified.

TABLE 10.1
Summary of Commercial Products (Commercial Adapted Eating Utensils)

Commercial Product	Source and Technical Characteristics	Price (US$)
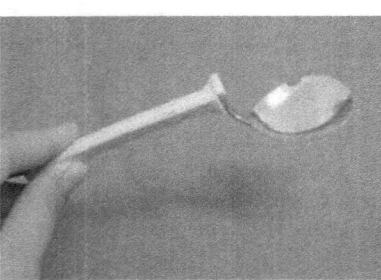	Balance-adapted spoon (Pachelli, 2014).	15.00 Cavalcanti, A. Information from the Hand Shop, obtained through a personal message received by <arturvaladares@gmail.com>, 10/6/2014.
	Covered spoon. Source: Assistivetech (s. d.). Spoon manufactured in two parts. It is made of plastic, ideal for smaller foods and liquids and easy to clean.	No information was found regarding the price of the product in Brazil, but abroad, the price/per unit is reported as $5.95.

(Continued)

TABLE 10.1 (*Continued*)

Commercial Product	Source and Technical Characteristics	Price (US$)
	Good Grips weight cutlery. Cutlery models with high weights, about 170 g, are indicated for people with PD with some type of spasticity or with some limitation of manual control. The handle is non-slip material, and the special metal shape of the cutlery allows it to be bent for better adaptation to the user. There is also a heavier model of about 227 g, suitable for people with greater motor difficulties.	The price is $8.95 per piece. WrightStuff.biz. Available at: <http://www. wrightstuff. biz/good-grips-eating-utensils. html>. Accessed: 10/30/2014.
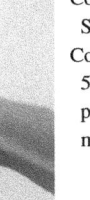	Commercial adapted spoon, Steady Spoon Phillips Mobility (s.d.). Commercial product from US 5630276 and US 5592744 patents, with counterweight-based mechanism.	Price per unit: $40.00.
	Liftware commercial adapted spoon (Liftware, s.d., 2015). Commercial product based on a fixed electronic apparatus that can be used with cutlery or different objects. The electronic system, powered by a rechargeable battery, can reduce tremors coming from the user's hand, by up to 70%. The device is, in a way, very similar to US patent 8308664 B2 in relation to the study presented by Pathak et al. (2014), whose main author is one of the inventors of the cited patent.	No information was found regarding the price of the product in Brazil; however, abroad, the price reported is $295.00.

The search for patents was carried out in four patent databases: Espacenet, INPI (Brazil), Google Patents and USPTO. Keywords with logical links were chosen for the search: for example, assistive device OR assistive technology OR assistive apparatus AND eating utensil OR spoon OR adapted spoon. Table 10.2 presents the better

TABLE 10.2
Identifying Improvement Opportunities of the Main Patents Related to the Problem

Description of the patent mechanism	Inspiration Conceptual design	Idea of the technical solution
US5592744A (1995)	Use of counterweight	Simple swivel system
US20100206885A1 (2010)	Adaptation to liquid and solid foods	Simpler format for food transfer

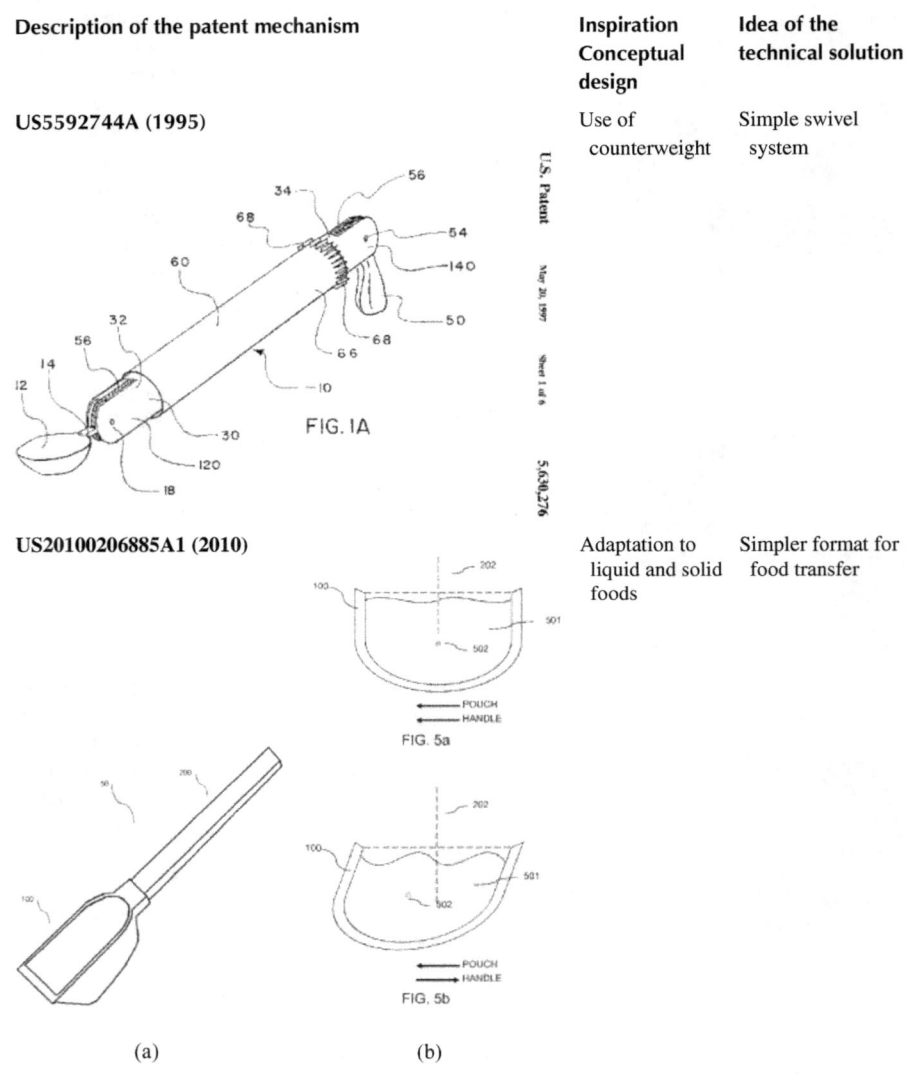

(a)   (b)

(*Continued*)

TABLE 10.2 (*Continued*)

Description of the patent mechanism	Inspiration Conceptual design	Idea of the technical solution
US8308664B2 (2010)	Interchangeable cutlery	Mechanical-only system
	Quick change of cutlery	Horizontal grip

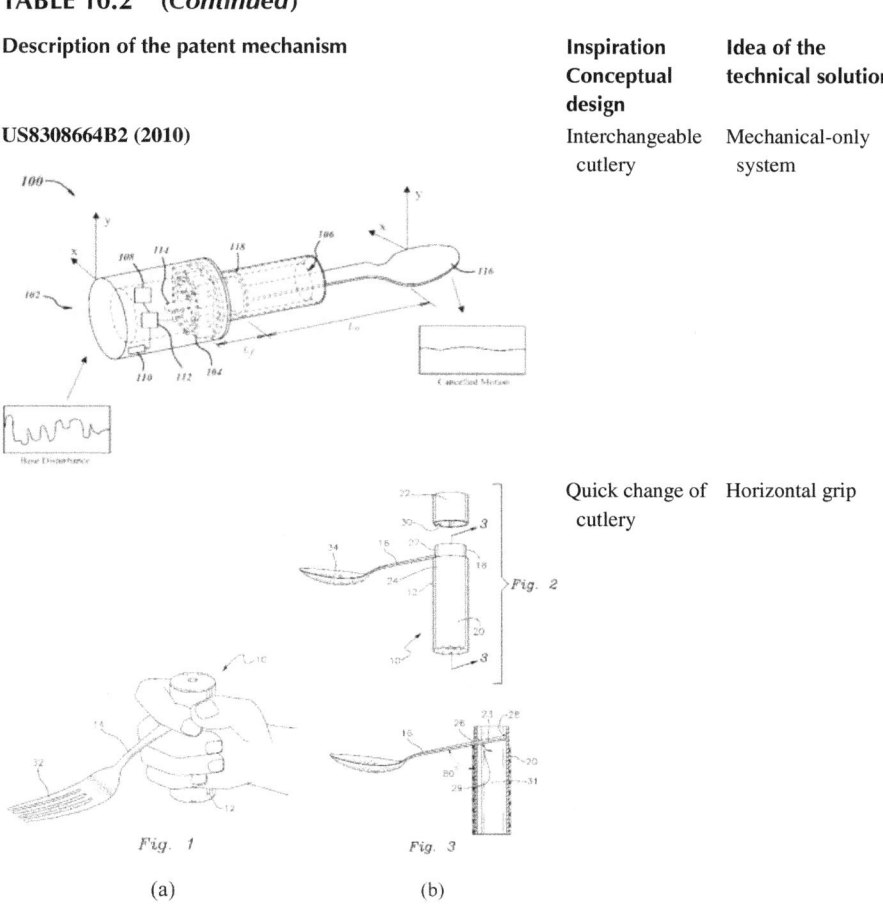

Fig. 1   Fig. 3

(a)   (b)

conceptual solutions extracted from the patent database considering the opinion of researchers in occupational therapy.

10.2.1.4 *Planning and Schedule*

The last task of the DCD phase refers to the understanding of the problem in the domain of interdisciplinary knowledge. Likewise, from the point of view of engineering, from the survey of the state of the art of commercial options, it is possible to identify design gaps in relation to the attributes required in assistive technology products, defined as: safety, comfort, independence and ergonomics. During this phase, a research project was developed, submitted and approved by the Ethics Committee of the Federal University of Triângulo Mineiro, Brazil (CAAE 35464114.4.0000.5154; Plataforma Brasil, saude.gov.br).

10.2.2 Conception Spiral

10.2.2.1 Design Requirements Definition

There are different techniques and approaches to obtain user requirements (short descriptions), such as in-person and online interviews, health tool assessment and open questionnaires, observation of activity user and video recordings. In this design phase, with interface concept generation, the integration of two design methods was chosen: House of Quality (HoQ) from quality function development (QFD) and the TRIZ technique, based on Chaoqun (2010) and Mayda and Borklu (2014), presented in Figure 10.3.

TRIZ is acronym of a Russian term, which in English means "The theory of inventive problem solving—TIPS". This design method is characterized by a systematic approach to innovation. Altshuller proposed the design method in the 1940s in order to generate a knowledge base from the patent database until this period. TRIZ structure is based on a set of inventive heuristics, called 40 inventive principles (IPs) and 39 engineering parameters (EPs). TRIZ then establishes the contradictions matrix (CM), a square matrix (40 × 39) that covers all the possible contradictions between pairs of these parameters. The design method proposes an iterative process that changes between these domains: from technical characteristics, it is possible to unfold into a set of EPs and then correlated IPs as part of the new problem solution. From this set of IPs, the conceptual solution can be explored by increasing the chance of innovation in the partial or total solution of the engineering problem. This process demonstrates that generation of innovative ideas does not occur in a random way but follows certain patterns and lines of evolution (Zlotin & Zusman, 2004).

Figure 10.4 shows a schematic synthesis of the process. Because it is often an abstract concept, the process of properly interpreting each IP in order to deploy it in technical solutions is very important. In this way, the search for examples, analogies and previous applications is very useful for a successful application. Thus,

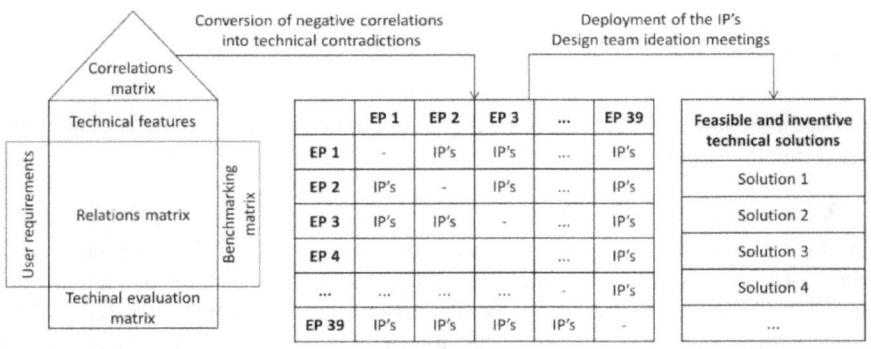

FIGURE 10.3   QFD and TRIZ integration.

Source: Santos et al., 2019. Adapted from Chaoqun (2010)

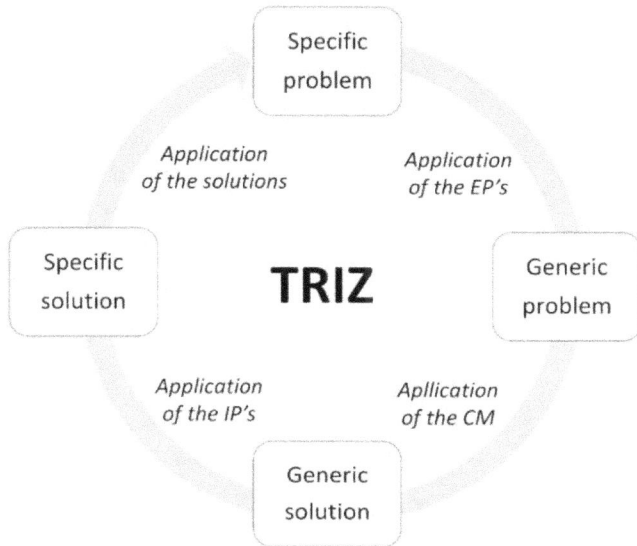

FIGURE 10.4   TRIZ process: iterative process between technical characteristics × EPs and IPs.

Source: Santos, 2015

TRIZ can be view as engineering problem-solving toolkit that gathers successful past technical solutions in order to support the elaboration of innovative solutions in a systematic way. In other words, TRIZ is a set of tools that help engineers understand and solve their problems by using the accumulated scientific and technological knowledge of the past (Gadd, 2011). Akao proposed QFD in the 1960s, and its objective, after the Second World War, was to offer a systematic design quality approach that could incorporate the user's requirements right at the beginning of design development, starting with the first matrix, called HoQ. User requirements are collected and refined on linguistic expressions, which are then translated into technical features. HoQ connects the technical domain and the user domain, transforming the importance associated with the user's set of requirements into relative weights corresponding to an often-expanded set of technical characteristics (Cheng & Melo Filho, 2007).

Users or stakeholders, who assign importance grades, evaluate the list of user requirements: one, three and nine, respectively: not very relevant, relevant and very relevant (Clausing, 1994). The design team established that each user could assign a maximum of two grades equal to nine, while the assignment of the others was free. The total grade for each requirement was then given by the sum of all grades received. For this case study, seven researchers filled out an open questionnaire, subsequently complemented by the COPM questionnaire, forming a three-iteration response and refinement process. These researchers also worked in the clinical area, after refining redundancies of expressions in a list of ten technical requirements, presented in Table 10.3, in ascending order.

TABLE 10.3
Level of Importance of User Requirements

User Requirements	Final Grade	Importance Degree
Compensate for tremor	51	Very Relevant
Be adaptable	51	Very Relevant
Do not hurt the patient	44	Relevant
Serve solid and liquid foods	39	Relevant
Be comfortable	39	Relevant
Be firm	39	Relevant
Support change of utensils	37	Relevant
Be easy to clean	33	Relevant
Low cost	25	Not Relevant
Present attractive design	13	Not Relevant

TABLE 10.4
Technical Characteristics (Technical Domain)

N°	Technical Characteristics	Magnitude
1	Hand-grip strength	N
2	Heat transfer rate at the interface	W/m^2
3	Preparation time for use	s
4	Number of Parkinson's degrees accounted	n
5	Number of pieces	n
6	Utensil capacity	m^3
7	Loss of food	%
8	Utensil cost	R$ (Real)
9	Mechanical strength	Pa
10	Number of edges	n
11	Cleaning time	s
12	Eating time	s
13	Vibration amplitude reduction	%
14	Frequency reduction	%
15	Inertia	kg
16	Volumetric compaction index	kg

Based on importance grade levels, the main user requirements were being adaptable (18%), compensating for tremors (15%), low cost (13%) and not harming the patient (12%). The diagonal trend of the matrix with strong correlation values represents a unique relationship between user requirements and technical characteristics. Based on Clausing (1994), these correlated values are represented by scale: 1—weak correlation, 3—average correlation and 9—strong correlation (practically, direct translation between user needs and technical characteristics). Table 10.4 presents a translated and converted set of technical characteristics.

It is important to emphasize that the technical characteristics, as defined by the House of Quality, must be measurable quantities. The definition of each of the technical characteristics is important for the conceptual stage itself. In this way, there are no doubts about the design functions that will be effectively studied or the integration of the roof of each quality, such as correlations in pairs of technical characteristics with TRIZ. After calculating the weights of the user requirements, it is possible to calculate the weights, absolute and relative, of the technical characteristics. The absolute weight of each characteristic is given by Equation (1):

$$PA_i^p = 100 \times \sum\nolimits_j User-Requirement\,relation/technical\,parameter_i \times PR_j^r \quad j \leq 10 \quad (1)$$

The relative weight, in turn, is given by Equation (2):

$$PR_i^p = \frac{PA_i^p}{\sum_j PA_j^p} \qquad j \leq 16 \qquad (2)$$

It is important to highlight that one of three possible values gives the user requirement/technical characteristic relationship in House of Quality: 1, 3 and 9. If there is no explicit relationship between the requirement and the parameter, the null value is adopted. The more important technical characteristics that represent the results of HoQ (engineering domain) were frequency reduction (14.29%), number of Parkinson's degrees considered (11.07%), vibration amplitude reduction (7.03%), utensil cost (8.61%), inertia (6.44%) and handgrip force (6.44%). The HOQ "roof", also called the correlation matrix, establishes the type of relationship between the technical parameters/characteristics of the product (Figure 10.5). In the correlation matrix (in pairs) obtained in the upper portion of the HoQ, a similar approach to the relationships between user requirements and technical characteristics can be adopted. These relationships can be of three natures: no relationship and weak and strong relationships, which are scored 1, 3 and 9, respectively, or by symbols that represent a scale for the pairs of contradictions defined as negative and positive contradictions. The positive signal indicates that one parameter generates an improvement in the other from a design perspective. Otherwise, the negative sign is assigned.

The six most relevant negative correlations, defined by the sum of the relative weights of the technical parameters involved, were converted into technical contradictions in TRIZ from the association of the technical parameters with the EPs.

As can be seen in Figure 10.5, the negative correlations are highlighted, as they are precisely those that most interest the design team as a starting point for their decision-making. In order to work more objectively in search of new technical solutions, an important criterion for negative correlations was established. The relative weights were added in the area the of technical evaluation matrix (Figure 10.5) of each of the parameters involved in a correlation, and from there the value of the relative weight of each correlation was obtained. The greater this weight, the more important the correlation for the design. The result of this process, containing the six most important correlation's, is presented in Table 10.5.

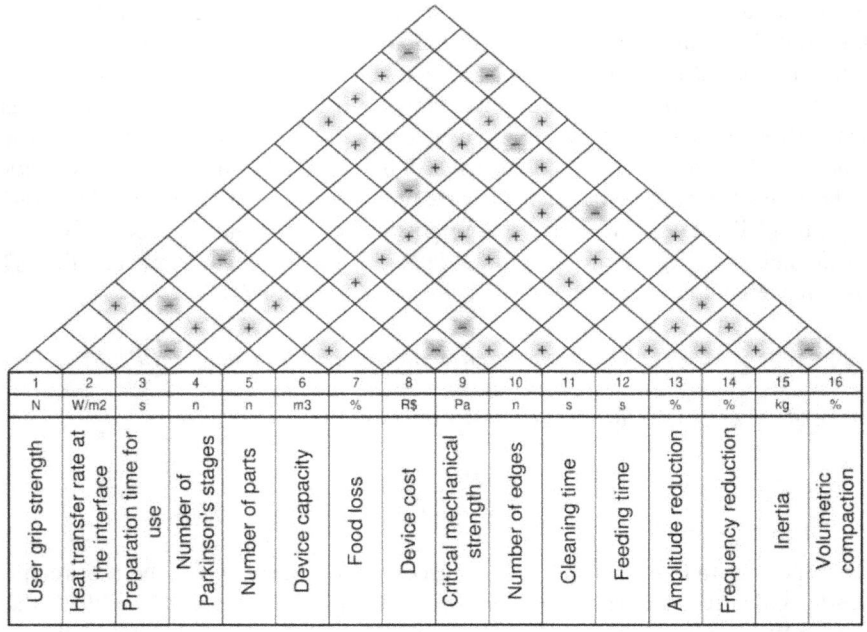

FIGURE 10.5 Correlation matrix (HOQ "roof") of the assistive device.

Source: Santos, 2015

TABLE 10.5
Relative Weight of Each Negative Correlation (Santos, 2015)

N°	Technical Parameter to be Improved	Technical Parameter to be Affected	Relative Weight (%)
1	Preparation time for use	Level of Parkinson's contemplated	16.99
2	Number of Parkinson's degrees contemplated	Eating time	15.64
3	Utensil cost	Inertia	15.05
4	Number of edges	Utensil cost	14.70
5	Inertia	Hand-grip strength	12.88
6	Utensil cost	Mechanical strength	12.18

In the following, the correlations between pairs of technical characteristics obtained from House of Quality are detailed to guide the application of TRIZ domains. Figure 10.4 presents the correlations matrix ("HoQ roof"), and Table 10.4 presents the most significant negative correlations based on relative importance resulting from the application of the first QFD (design domain) or matrix HoQ.

Correlation 1: the shorter the fixing time of the utensil to the user's hand (without the aid of a third party), the better the condition of the first parameter. For the most severe Parkinson's degrees, however, this time has a reduction limit, due to the user's

TABLE 10.6
TRIZ Final Results (Santos, 2015)

Negative Correlations between Technical Characteristics (QFD)	Technical Contradictions between EPs (TRIZ)
Preparation time for use	33—Ease of operation
x	x
Number of PD stages contemplated	35—Adaptability
Number of PD stages contemplated	35—Adaptability
x	x
Eating time	39—Productivity
Apparatus cost	36—Device complexity
x	x
Inertia	35—Adaptability
Number of edges	12—Shape
x	x
Apparatus cost	29—Manufacturing precision
Inertia	35—Adaptability
x	x
User grip strength	10—Force
Apparatus cost	32—Ease of manufacture
x	x
Critical mechanical strength	14—Strength

own motor difficulties. Thus, an improvement in the first parameter comes up against a limit imposed by the second and may even harm the latter.

Correlation 2: an improvement in the first parameter means that the tool will also be able to meet the most severe degrees of Parkinson's, which in turn require more time to complete the meal, thus establishing a limitation to the second parameter.

Correlation 3: the reduction in the total cost of the tool generates an improvement in the first parameter. This, however, generates a limitation to the second parameter, since an increase in mass, up to a certain limit, or a mechanism that allows its variation, represents an improvement in the second parameter. Thus, a negative correlation emerges.

Correlation 4: a reduction in the number of product edges means an improvement in the first parameter. However, this reduction increases the utensil's manufacturing costs, generating a loss in the second parameter.

Correlation 5: up to a certain limit, the increase in the inertia of the tool represents an improvement in the first parameter. This increase, however, generates an increase in the user's handgrip strength and, therefore, a loss in the second parameter.

Correlation 6: an improvement in the first parameter represents a reduction in the total cost of the product. This, however, means, at least in part, the use of less material or material of lesser quality and/or resistance and, therefore, damage to the second parameter. Table 10.6 presents the final EPs obtained from application TRIZ process.

Table 10.6 presents the result of TRIZ application of technical characteristics—pair comparison.

Figure 10.6 presents the sketches to propose a technical solution from the TRIZ results.

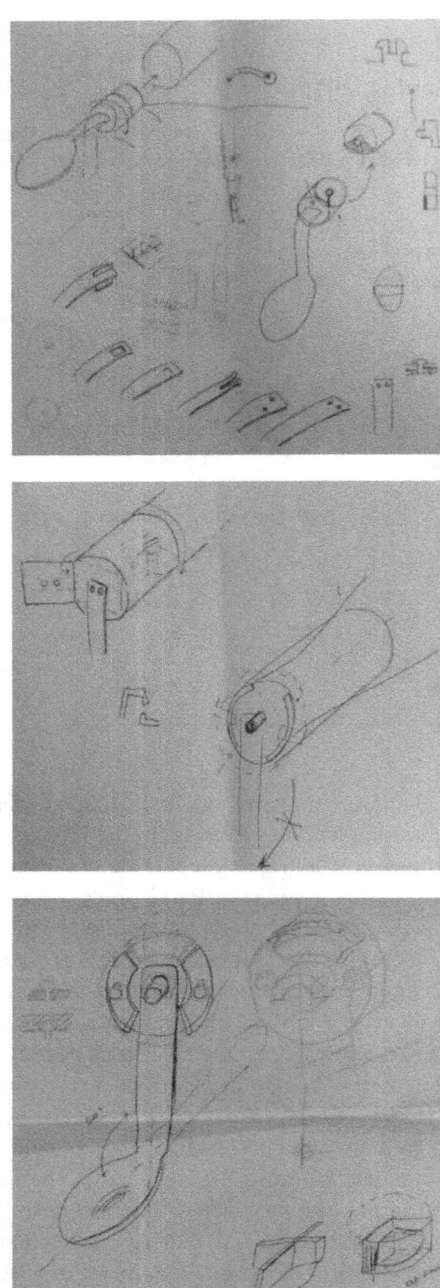

FIGURE 10.6   Sketches generated after TRIZ application for conceptual solution for assistive eating device.

Source: Santos, 2015

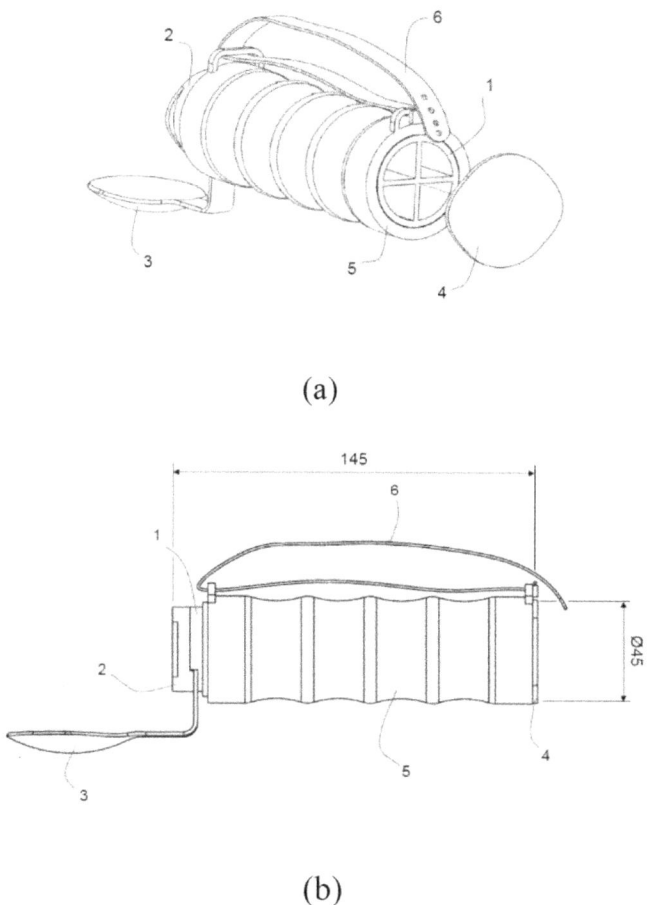

FIGURE 10.7 Assistive eating device: (a) technical drawing (isometric view); (b) front view with main dimensions (guided in advance by occupational therapist).

Source: Santos et al., 2016

Finally, the main innovative technical characteristics for the concept of the adapted assistive device for self-feeding were: passive mechanism, ease of use, replacement and maintenance of components; use of a non-toxic fluid (water) that is easy to transport and replace, with the aim of helping balance by minimizing tremors for people with a milder degree of Parkinson's disease; easy assembly/disassembly; and larger area of manual contact, providing greater grip strength and greater thermal and ergonomic comfort for the user. According to healthcare researchers, the conceptual solution and functional prototype met the users' requirements, providing adaptability and the possibility of customization at a relatively low cost (about US$10.00).

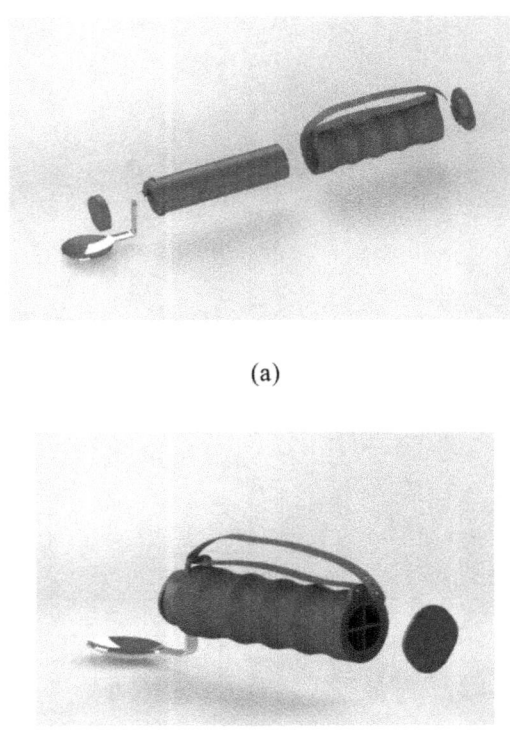

(a)

(b)

FIGURE 10.8 Virtual model of the conceptual solution for assistive eating device.

The technical drawing is shown in Figure 10.8. The AD for self-feeding is composed of: main handle (1) to which the other parts are attached: adapted cutlery (3), a lid to hold the cutlery (2), a rear sealing lid (4), an ergonomic cover (5) and an adjustable strap (6). The cable also contains four internal cavities, which present filling level variations in number and volume, totaling five possible weight configurations for the device until all cavities are completely filled.

10.2.2.2 VIRTUAL MODEL CREATION

The geometric models naturally used in engineering are highlighted by researchers in frontier areas, mainly due to the properties and resources of the technical drawing. Physical properties such as texture and kinematics resources help to better visualize the concept. Researchers in the engineering area begin the process of manufacturing adequacy and verifying gaps and geometric errors, which will have an impact on assembly. Figure 10.8 presents the geometric model developed in a CAD system.

10.2.2.3 *Material Selection and Choice of Manufacturing Process*

A low-cost additive process was the first choice for fabrication of the conceptual prototype. Two main reasons supported this option: the injection process at this stage would be infeasible due to costs and time, and the design would still have some changes. This design phase and customization design approach highlight the importance of design democratization, including open manufacturing with easy access within the academic environment. Poly-lactic acid (PLA), a synthetic thermoplastic polymer that is biodegradable and has non-toxic properties was initially chosen for fabrication of the adaptive eating device due to easy access as feedstock for 3D printers based on fused filament fabrication (FFF). Despite their inherent degradation, the design team and occupational therapists in a closed environment evaluated the prototypes. In the first version, all parts, including the spoon, were printed using PLA. In the following versions, conventional spoons were used, with better finishing at the edges to be effectively used in testing with the end user. The adaptation of these commercial spoons to later versions included a process of cutting, bending, drilling and polishing to fit the assistive device.

10.3 EVALUATION AND REFINEMENT PHASES

After process planning for additive manufacturing of separate parts, the first version of the adaptive eating device (mock-up) is presented in Figure 10.9. This conceptual prototype was used to present the technical solution to the clinical research in occupational therapists in order to obtain feedback on the design's assistive technology attributes. After some preliminary adjustments, the second version of the prototype was obtained—the first functional one—which can be seen in Figure 10.10.

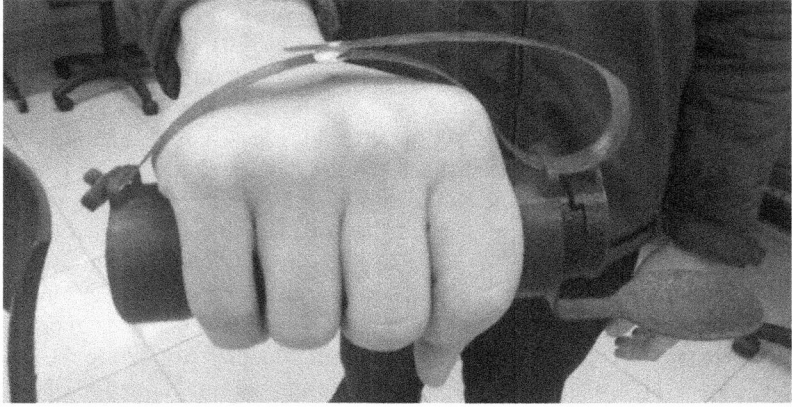

FIGURE 10.9 Mock-up of the first version of the adapted eating device.

Source: Santos, 2015

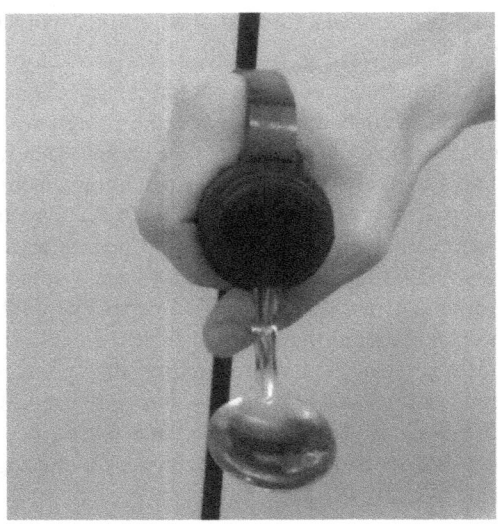

FIGURE 10.10   Second version of the conceptual prototype of the adapted eating device.

Source: Author, 2016

Figure 10.10 shows the use of an adapted commercial spoon (stainless steel), where the design team decided to use this second version with some modifications for the clinical tests with the single-user method.

10.3.1 Third Version of the Prototype

At this stage of technical refinements of the conceptual solution, the end user and occupational therapists (OTs) evaluated the second version of the device. A 60-year-old man with Parkinson's disease with regular use of his medications was selected to carry out the AD assessment tests. In Hoehn-Yahr modified scale, from 1967, its condition was classified as stage 2 or 2.5 (mild or moderate) (Vieregge et al., 1997). This clinical procedure was supported by a research project approved by the Ethics Committee of the Federal University of Triângulo Mineiro, Brazil (CAAE 35464114.4.0000.5154), where the participant signed the informed consent form. The tests presented in Figure 10.11, preliminary and more subjective, aimed to listen to the user's experience and opinions regarding the product during use in the feeding task, with monitoring by OTs. The purpose was to identify possible adjustments, improvements and/or modifications to be made in the next version of the prototype.

It is important to report that, for the most part, at a distance, while the user and OTs performed the tests in Uberaba, MG, the design team remained in São Carlos, SP. For this reason, it was necessary to use the means of communication and media available today (email, messages, audio and, in particular, video) to exchange information, monitor the progress of tests and observe functional improvements. Some visits were also made to Uberaba to monitor the process and conduct interviews

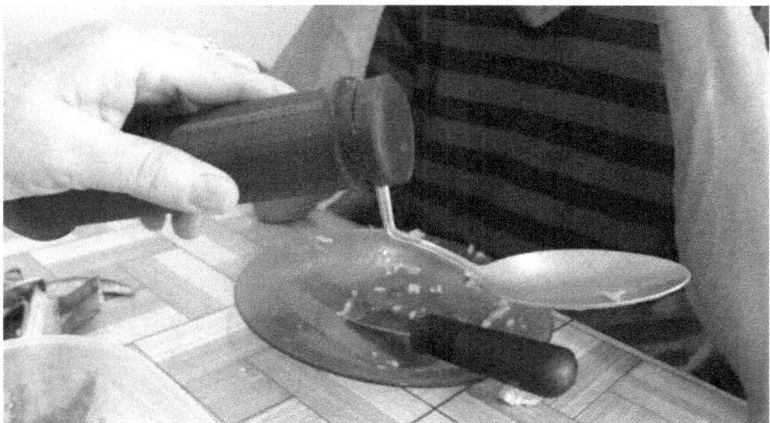

FIGURE 10.11   First tests with the second version of the conceptual prototype of the adapted eating device.

Source: Author, 2017

with the end user and family members. For these first tests, the mechanism for fixing the front cable cover—designed to be of the quick-fit type (snap fit)—was not yet well adjusted, and therefore a temporary solution was adopted with the use of a temporary adhesive (Blu-Tack) for fixing. The result was not satisfactory, and the lid came loose with the movement of the cutlery. Therefore, glue was used for the new fixation in such a way as to allow the rotary movement of the cutlery. Even so, the configuration of the parts and the fixation caused frequent locking of this

(a)   (b)

FIGURE 10.12 Modifications in the front of the cable: (a) the front cover; (b) from the second (top) to the third (bottom) version of the prototype.

Source: Author, 2017

movement. In addition, the material of which the product was manufactured was not flexible enough to adopt the quick coupling strategy as previously thought. The tabs for removing the front cover were not suitable, because they were too short, as reported by the user and OTs.

These factors led the design team to modify the front part of the cable and the front cover for the next version of the prototype, changing the fastening process but maintaining a quick coupling strategy. The taps were also modified, being made wider and with a hexagonal configuration to facilitate handling. A comparison between the two versions is presented in Figure 10.12. It is important to note that in this new configuration, it was necessary to carry out a study of the clearances/tolerances due to the additive process (low-cost 3D printing equipment and variations in the geometry) during and after the manufacturing process in order to obtain an adequate fit between the parts. For this, several parts were printed with small variations in diameters in search of the most appropriate fit.

Another example of clearance setting in the additive manufacturing process can be seen in Figure 10.13. Taking the cable as a reference, three small sections of the cover were produced, with slight variations in the internal diameter—from the

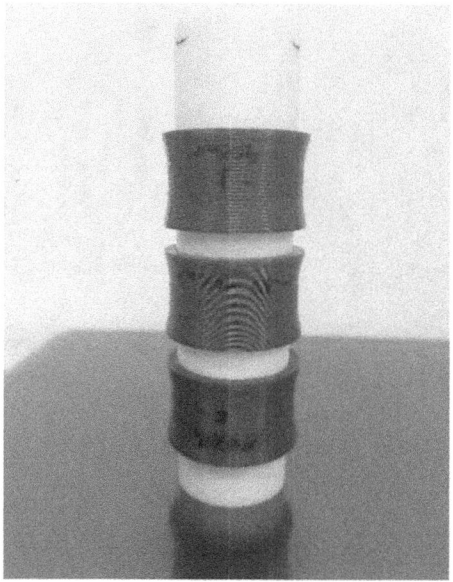

FIGURE 10.13   Clearance fitting between cable and cover.

Source: Author, 2017

nominal diameter in the drawing—in order to find the best fit. This same adjustment value in reference to the nominal value was then used for the other cylindrical part fittings.

In relation to the rear cover, whose fixation had also been designed to be fast coupling (snap fit) by pressure, the first alternative, of temporary character, used adhesive (Blu-Tack) for fixation. For clinical testing without using water in the internal cavities of the cable, there were no problems. However, in the tests with one or more of the cavities filled with water, the seal was not satisfactory, with significant loss of the volume of water deposited and, therefore, with damage to the task of eating and the desired functionality for the eating device adapted with the extra weight. As a second option, still of a temporary nature, thread sealing tape was used, which significantly improved both the sealing and the fixation (by pressure).

As in the case of the front cover, the configuration of the support taps for removing the rear cover was also described as unsatisfactory by the end user and OTs. From these observations, the cover for the third version of the prototype had some more modifications presented in Figure 10.14: change in the cover removal tabs, a hexagonal configuration to facilitate handling and a slight increase in the thickness of the fixing lugs for better fixation of the thread seal tape.

Design team members proposed other small technical improvements: despite the improvements in sealing provided by the previous changes and the use of thread sealing tape, which already allowed for clinical testing to be carried out with the cavities filled, the solution was still not completely satisfactory, with some leakage

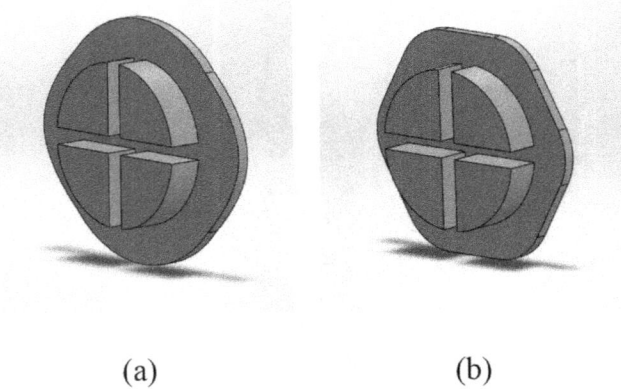

(a)   (b)

FIGURE 10.14   Modifications in the rear cover from (a) the second to (b) the third version of the prototype.

Source: Author, 2017

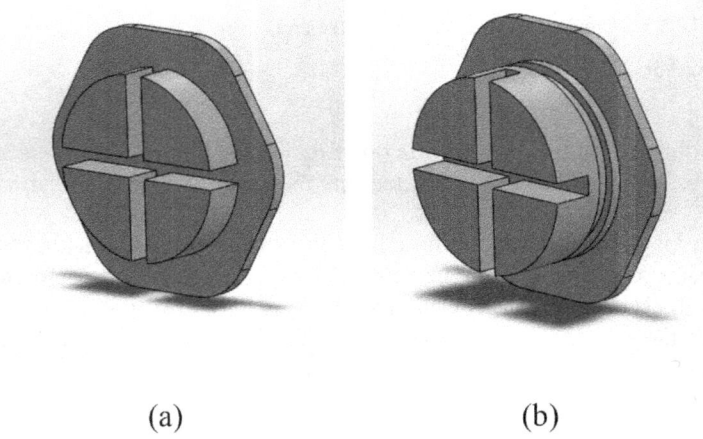

(a)   (b)

FIGURE 10.15   Changes to the rear cover from (a) the third to (b) the fourth version of the prototype.

Source: Author, 2018

during continued use of the assistive device. For this reason, a new cover model was developed, with thicker fixing lugs and the insertion of another lug for placing an O-ring. A comparison between the models can be seen in Figure 10.15.

There was also a modification in the metal cutlery used for the adapted assistive device for self-feeding. The model used in the second version of the prototype was considered very light, compromising in part the rotational movement expected for balance of the food. In addition, the quality of the finish was reported as poor by the

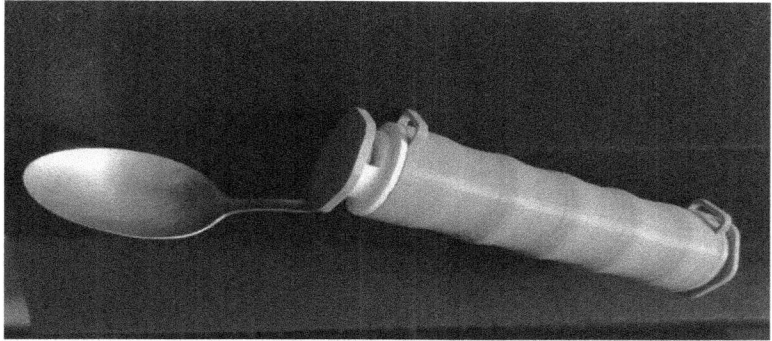

FIGURE 10.16   Third version of the assistive device for self-feeding (functional prototype).

Source: Author, 2017

user, which caused discomfort in the task of eating. Copolymer acrylonitrile buta-diene styrene (ABS) was also another change to manufacturing of the third version of the prototype. This change occurred due to the low degradation of the material; better dimensional control; easier surface finishing through mechanical sanding; and chemical reaction with acetone, which gives greater superficial quality. The third version of the prototype (now functional prototyping), with the modifications men-tioned previously, can be seen in Figure 10.16. This version was also tested with the end user and OTs. Up to that stage, the prototype had been used without the adjust-able belt.

10.3.2   Fourth Version of the Prototype

In this phase of the project, a one-year academic agreement was signed between the University of Sao Paulo (São Carlos Engineering School) and a local start-up (Model Works LTDA). The objective of this partnership was to make feasible a usability evaluation with a potential end user. From this collaboration, some important mod-ifications to the design were made. The first one refers to the thickness of the lids, which was increased on both the front and the back cover to facilitate their handling. In the case of the rear cover, a taper was also inserted at the end of the fixing lugs to facilitate its introduction into the cable cavities, as shown in Figure 10.17.

The front part of the handle and the front cover, which restrict the longitudinal movement of the cutlery, also received modifications to give greater stability to the tilting movement of the cutlery, reducing friction and preventing possible mechanical locking. In addition, several edges of the prototype were rounded in order to smooth the end user's contact with the parts and reduce stress concentrations. The compar-ison between the new and previous versions of the front of the cable and the covers can be seen in Figure 10.18. Another important modification of the fourth prototype in relation to the third was in the manufacturing and post-processing. In order to pre-vent a very long post-processing time due to the removal of the support material, the cable was now manufactured in separate parts in the 3D printer. So, small identical

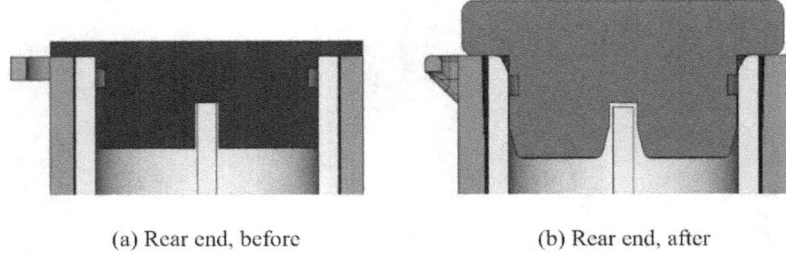

(a) Rear end, before   (b) Rear end, after

FIGURE 10.17   Modifications in the rear cover in the fourth version of the prototype.

Source: Adapted from Mattazio, 2019

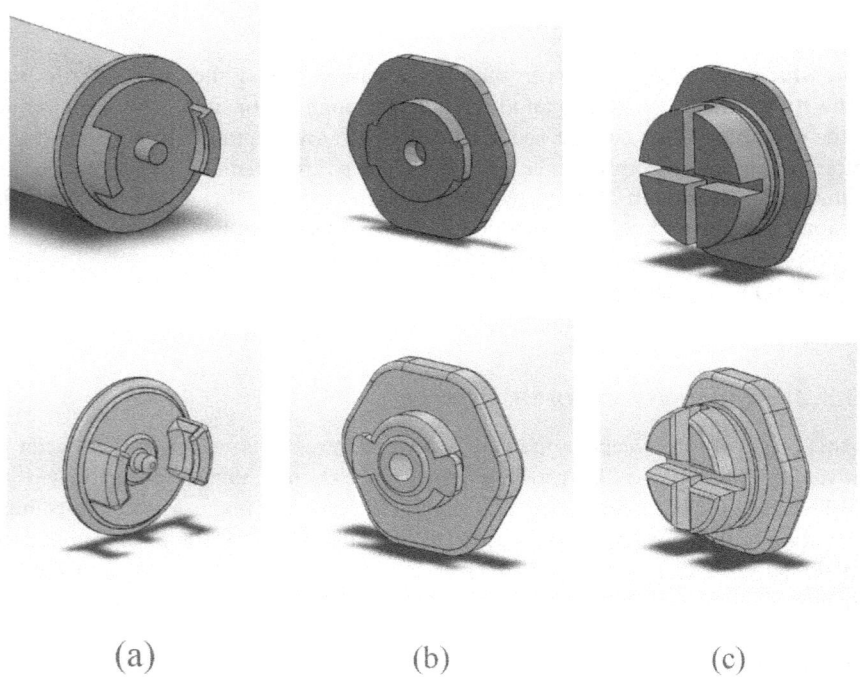

(a)   (b)   (c)

FIGURE 10.18   Modifications in (a) the front part of the cable and in the (b) front and (c) rear covers from the third (top) to fourth (bottom) version of the prototype.

Source: Author, 2019

recesses of square shape were added to the pieces to be joined, as can be seen in Figure 10.18. Small pieces (inserts) were also manufactured in this same shape, with intermediate thickness—greater than the depth of the shoulder but less than the sum of both—so that they could be used as a connecting element in the fitting of the two

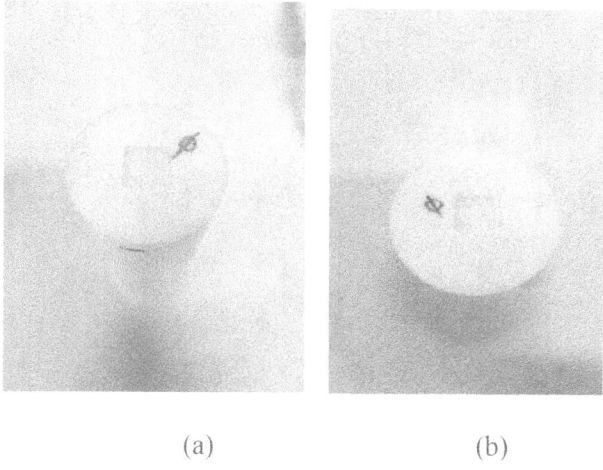

(a)   (b)

FIGURE 10.19   Cable: (a) front of cable, (b) with slot for connection.

Source: Author, 2019

parts, as shown in the schematic technical drawing presented in Figure 10.19. The objective was to facilitate and guide the join of the pieces, maintaining the proper alignment between the parts. Finally, the union was consolidated through bonding (PVC glue).

A "French hand"–type support was also added to the fixing strap on the cover, with the objective of increasing the mechanical resistance and facilitating the post-processing of the part, thus avoiding the insertion of support material along the entire part in the axial direction in the additive process. Regarding the belt, still absent in the third version of the prototype, a commercial model of silicone with metal clips was adopted in this new version. These differences between the versions can be seen in Figures 10.19 and 10.20.

Changes in the cover belt can be observed in Figure 10.21. These technical modifications led all parts to undergo adjustment of 3D printing tolerances at the interfaces between them. Regarding the thickness of the layer manufactured using the FFF additive process, one that represents a balance between the mechanical strength and time of manufacture was selected. In the case of cables, whose internal cavities can be filled with water during the use of the functional prototype, a test was carried out to verify if the manufacturing parameters (filling, layer thickness) were sufficient to keep them watertight, preventing water leakage. For this, some DAs were completely filled with water and left in an upright position for a few hours, as shown in Figure 10.22. The sealing efficiency provided by the new configuration of the cover together with the O-ring was also verified with this test. First, the infiltration of water inside the part was identified after some time in the indicated position, a problem that was solved by modifying the manufacturing parameters for the next parts.

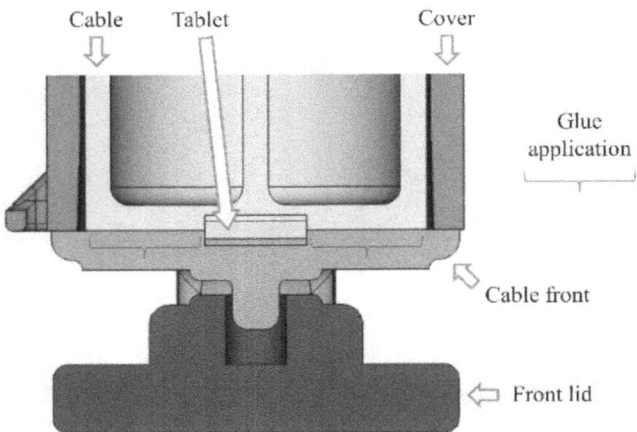

FIGURE 10.20 Schematic technical drawing of the join of the front part of the cable by an insert.

Source: Adapted from Mattazio, 2019

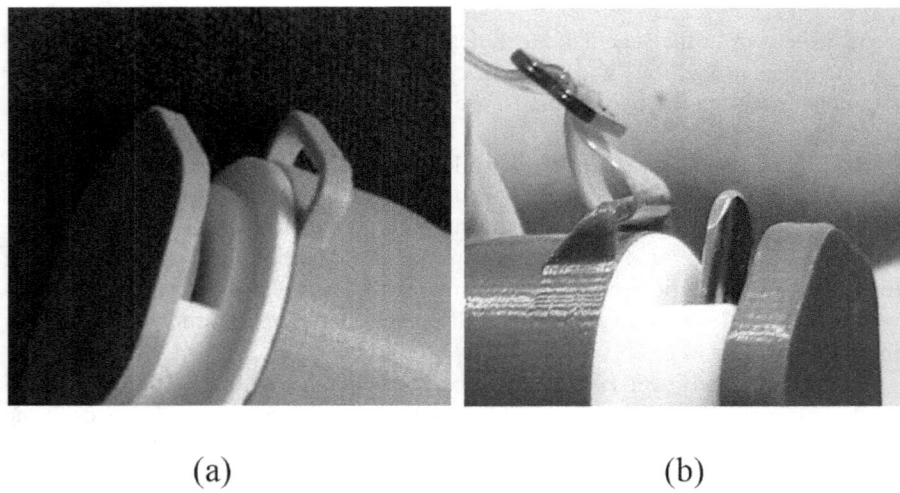

(a)   (b)

FIGURE 10.21 Changes in the cover belt from (a) the third to (b) the fourth version of the functional prototype.

Source: Santos & Siveira, 2020

At the suggestion of the end user, who sometimes complained about the unfavorable angle at which the cutlery reached his mouth, an alternative version of the prototype in which the cutlery had an "inward" curvature (Figures 10.22 and 10.23)

FIGURE 10.22 Adaptive device for self-feeding.

towards the user was designed for comparative tests, ceasing to be aligned with the longitudinal axis of the assistive device, and the comparison between the two versions, with straight and curved cutlery, is shown next.

Figure 10.24 shows the fourth version of the prototype with the modifications mentioned previously, in its complete configuration for use, with all parts assembled. Figure 10.25 presents the four versions of the prototype: (1) and (2) conceptual prototypes and (3) and (4) functional prototypes.

In summary, the main technical changes in the device focused on resolving three factors: use of the O ring in the radial position (shaft hole), because there was no device that would generate axial force, and seat dimensions according to the technical catalog. The use of the FFF additive process involved more than one iteration to adjust the interference in order to balance the sealing and ease of placement for the user and occupational therapist, and, finally, the surface finish of the O-ring region had to be the best possible to favor sealing, which involved experimental planning in printing and later sanding.

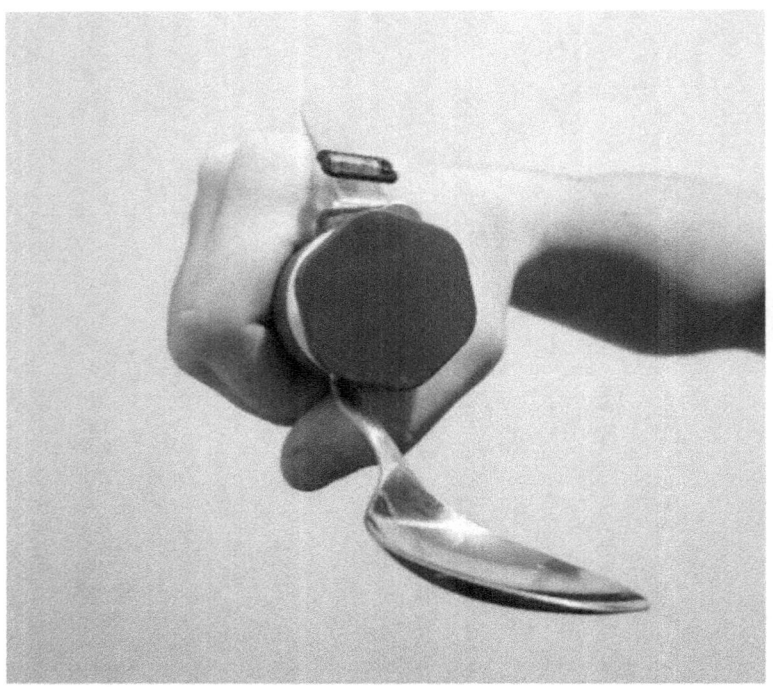

FIGURE 10.23   Alternative adapted eating device version with curved cutlery.

Source: Santos et al., 2017

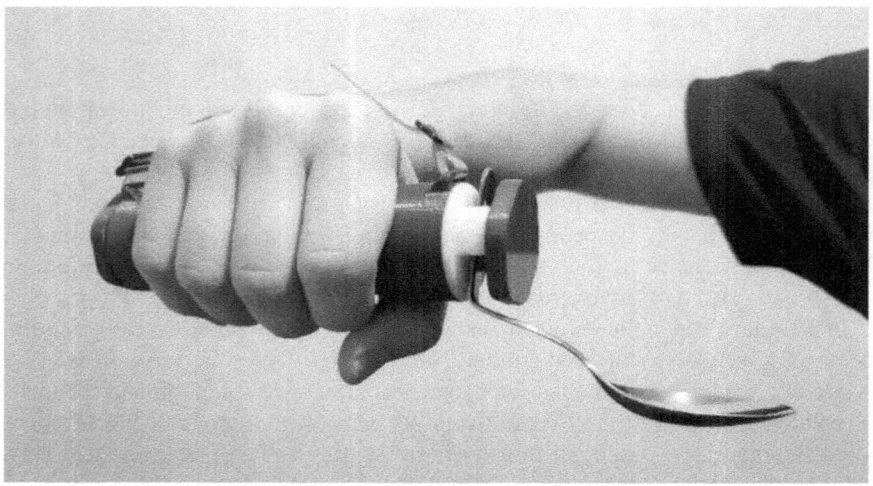

FIGURE 10.24   Fourth version of the adaptive device for self-feeding (functional prototype).

Source: Santos, 2018

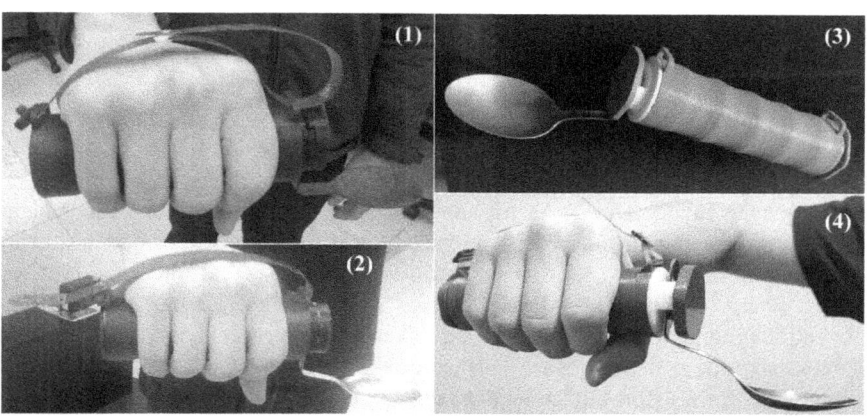

FIGURE 10.25 Comparison of the four versions of the prototype.

Source: Santos et al., 2017

10.4 CLINICAL TESTING: SINGLE-USER METHOD

Simultaneously with the fourth prototype improvement process, the third version was submitted to single-subject tests for functional evaluation of the prototype, with the same participant as the preliminary tests. This step was again supported by the research project approved by the Ethics Committee of the Federal University of Triângulo Mineiro, Brazil (CAAE 35464114.4.0000.5154). A single-subject statistical (ABC type) study was selected (Portney & Watkins, 2015). This statistical treatment can be very suitable for occupational performance assessments, because it allows evaluating the effects of one or more specific interventions under controlled conditions for a given individual. Each participant therefore acts as their own control group, and interventions are applied systematically, with repeated measures over time of the same dependent variable (or variables) in a period of no intervention (baseline) followed by a period of intervention (DEITZ, 2006).

Occupational performance and the degree of satisfaction were measured with the support of the COPM. The objective was to evaluate and identify changes over time in the perception of the end user about their performance in the occupations of their daily routine. It is noteworthy that the COPM is an assessment instrument in the health area obtained through semi-structured interviews that has been translated and validated for the Brazilian context (Law et al., 2009). Based on Law et al.'s (2009) methodology, only steps 4, 5 and 6 were considered. Step 4 is characterized by obtaining a performance score in the eating task before the adoption of the adapted eating device, where the end user assigns a score in the range of 1 (unable to do) to 10 (able to do extremely well). In the next step, 5, the end user assigns a score for their satisfaction in relation to the feeding task before the adoption of the adapted eating device, on a scale between 1 (not satisfied) and 10 (extremely satisfied). Finally, in step 6, a re-evaluation of the scores for performance and satisfaction is done after the intervention with the use of the AD in order to identify possible changes. The participant in the single-subject study was the same as in the preliminary tests.

Occupational performance data were collected over 70 days, with the exception of weekends and holidays, between April and August 2017. During this period, the user's perception of their performance in relation to the feeding task and their satisfaction with it were evaluated with the aid of the COPM, which was applied to each visit by the OT after the task had been performed. The performance score was recorded after the following question: "What grade would you give to your performance in the task of eating with this cutlery?" As for satisfaction, the question was: "How satisfied are you eating with this cutlery?" The variables of interest were evaluated in three distinct phases: baseline (A), first intervention (B) and second intervention (C). In the first phase (A), lasting 21 days, variables were verified in relation to the use of conventional cutlery. In phase B, the same variables were analyzed for a period of 31 days using the assistive device developed with the cavities without water. Finally, for phase C, the assistive device with a cavity filled with water was used for 18 days.

The data was analyzed by support of three methods: (1) celeration line or trend line, (2) band with two standard deviations and (3) visual analysis. The first one was based on a reference line that expresses the general trend of the obtained data, plotted with the support of the split-middle technique, in which the baseline phase is divided into two symmetrical parts, so that the same amount of data is represented above and below the trend line. The null hypothesis assumes there is no change in the trend line between one phase and another; that is, when extending the trend line to the intervention phase, the data would also be divided equally. Statistical significance was determined by the binomial test ($a = 0.05$) (Portney & Watkins, 2015). The two-band standard deviation (SDB) method is based on analysis of the variability of phase data over baseline. The mean and standard deviation of this phase are calculated and, with this, two lines are drawn that delimit the range, with two standard deviations above and below the mean. These lines are extrapolated to the intervention phases. Changes between the intervention phases and baseline can be considered significant if at least two consecutive points of the intervention phase are outside the band (Portney & Watkins, 2015).

Visual analysis was used as support for the final evaluation of the results, mainly in cases where there was disagreement between the results of the two previous methods (speed line and two standard deviation range). This analysis was based on three main characteristics: data level, trend or direction of change and angle of inclination presented in each phase (Portney & Watkins, 2015). Regarding the level, when there is a difference between the last data point of a phase and the first of the next adjacent phase, it is considered that there was a change in level between them. When the data have a certain stability and low angle of inclination in the trend chart, the comparison of the data means between two phases is also another means of assessing the level change. Regarding the trend, the data may be accelerating, decelerating or stable in a linear or curvilinear manner. Finally, as for the angle of inclination, it can be greater or lesser in one phase in relation to the other, indicating a greater or lesser rate of change (Portney & Watkins, 2015).

The results presented in Figure 10.27 indicate the performance function evaluations. The mean score for the phase in relation to the baseline was 6.19 (standard deviation [SD] = 1.12), with a minimum score of 4 and a maximum of 8 points. In phase B, with intervention, the mean was 6.87 (SD = 1.28), with scores between the

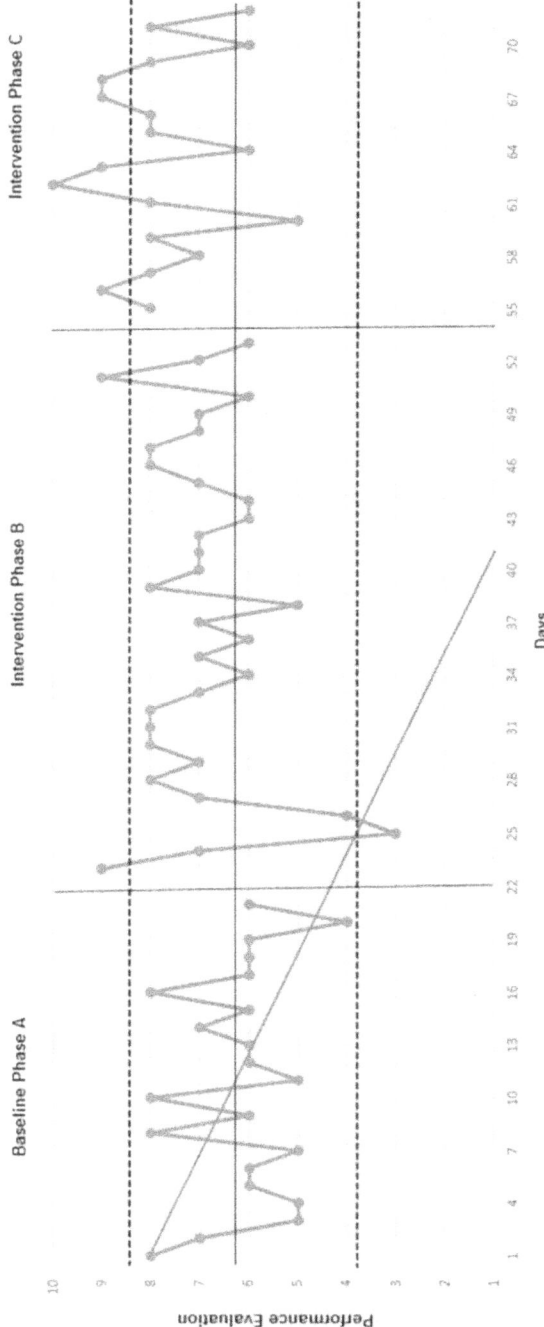

FIGURE 10.26 Results for the performance variable over time, supported by COPM.

Source: Adapted from Cavalcanti et al., 2020

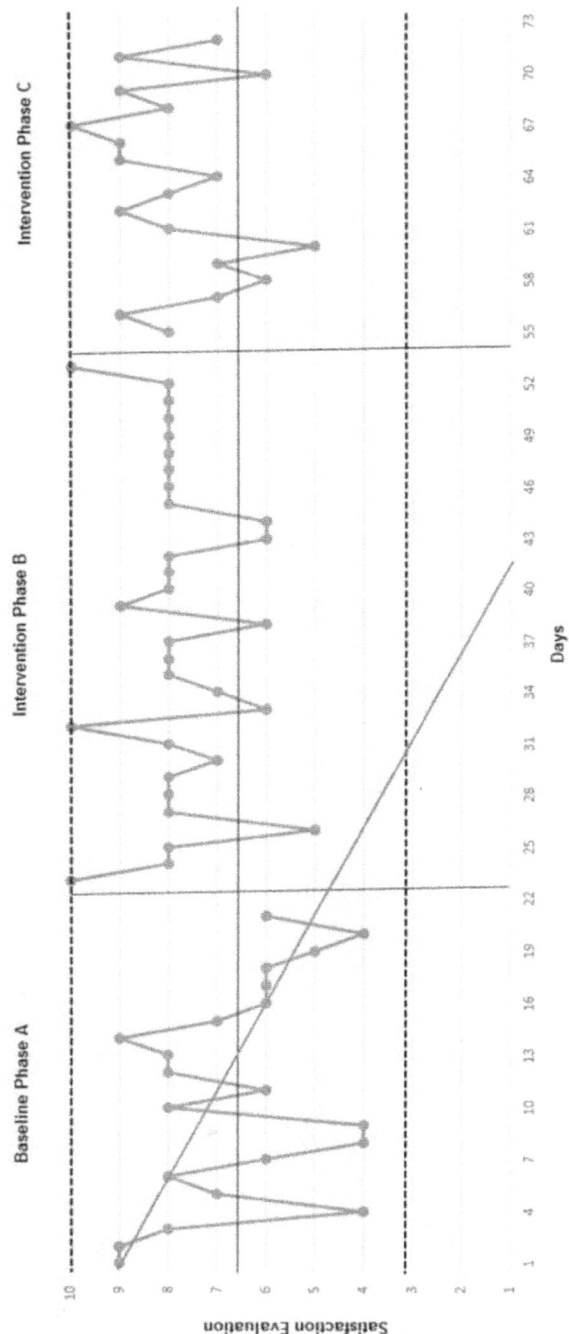

FIGURE 10.27   Results obtained with the aid of COPM for the satisfaction variable.

Source: Adapted from Cavalcanti et al., 2020

extremes of 3 and 10 points. Finally, in phase C, also with intervention, the mean was 7.77 (SD = 1.31), ranging between 5 and 10 points. The celeration line indicated a deceleration trend in phase A (baseline), while in intervention phases B and C, it was possible to perceive an acceleration trend, with more points above the trend line. The binomial test indicated statistical significance ($p < 0.05$) when comparing the two phases of the intervention in relation to phase A, emphasizing the trend observed in the celeration line. Based on the two-standard deviation band method, the variability of the data and the absence of consecutive points outside the band indicated that there was no statistical significance in phase B. In view of the disagreement between the two analyses of the variable, visual analysis was used. This analysis indicated a change in the level between phases (A and B) based on the comparison between the last point of the first (smaller) and the first point of the second.

The difference between the averages of the phases also showed an increase in the level of performance of phase B in relation to the baseline. The presence of 22 points above the baseline midline in phase B also indicates a direction of change for celeration line. The results of the visual analysis corroborate those of the celeration line, suggesting an increase in the performance of phase B in relation to phase A. In the case of phase C, it can be observed by the method of the band of two standard deviations that there was an occurrence of consecutive points outside the band, despite the variability of the data, thus suggesting a statistical significance in the change of the performance variable in this phase, reaffirming the perception obtained by the celeration line.

Table 10.7 presents a summary of the analysis using each method and the results obtained.

Figure 10.28 shows the results of the satisfaction ratings. The average was 6.57 (SD = 1.72), with a minimum score of 4 and a maximum of 9 points. In phase B of the intervention, the mean was 7.81 (SD = 1.14), ranging between 5 and 10 points. Finally, in intervention phase C, an average of 7.83 (SD = 1.34) was obtained, with a score between 5 and 10 points. Based on the celeration line, a deceleration trend was identified in phase A. On the other hand, in phases B and C, the higher proportion of points above the trend line indicates an acceleration trend, with some periods of the celeration line. The binomial test showed statistical significance in the two intervention phases ($p < 0.05$), reinforcing that observed through the celeration line.

The two-band standard deviation method indicated, in addition to a large variability of points in the intervention phases (mainly in phase C), the absence of two

---

TABLE 10.7

Results of the Analysis Methods for the Performance Variable. Adapted from Cavalcanti et al. (2020)

Performance	Celeration Line	Two Standard Deviation Band	Visual Analysis
Phase B	S ($p = 1,44$E-8)	NS	S
Phase C	S ($p = 3,81$E-6)	S	—

($p < 0,05$; S = statistically significant result; NS = non-significant result)

TABLE 10.8

Result of the Analysis Methods for the Satisfaction Variable. Adapted from Cavalcanti et al. (2020)

Satisfaction	Celeration Line	Two Standard Deviation Band	Visual Analysis
Phase B	S	NS	S
Phase C	S	NS	S

($p < 0{,}05$; S = statistically significant result; NS = non-significant result)

consecutive points outside the band, indicating the lack of statistical significance in the change in the variable in these two phases. As there was disagreement with the results obtained with the speed line for these phases, visual analysis was used again. Visual analysis indicated that changes within the levels can be identified between the intervention phases—phase B (mean = 7.81) and phase C (mean = 7.83)—and baseline (mean = 6.57). The fact that the last point of phase A is smaller than the first value of both phases B and C reinforces the perception of the change in level between them.

The presence of 24 points above the baseline mean in phase B and 11 points in phase C also indicates a change in the direction of change to the celeration line, despite the presence of opposite points. These results are similar to the results obtained by the celeration line, suggesting an increase in satisfaction in the intervention phases in relation to phase A.

Table 10.8 summarizes the results obtained by each three methods.

The statistical results indicated that the use of AD for self-feeding improved the user's occupational performance and generated satisfaction compared to conventional cutlery. The performance results also indicated that there were functional gains in the self-feeding task, a condition that reinforced user satisfaction with the use of AD. As general conclusions of the evaluation by the single-user method on the developed AD; the diameter of the cable with the largest contact area for gripping; the use of polymeric material; and the semi-alternating movement of the interchangeable tip, even without filling the cavities with liquid, presented better occupational performance and user satisfaction in the self-feeding process. The use of a cavity filled with water did not show any significant difference in the results.

The proposed design materialized in the functional prototype presented significant improvements in the AD for self-feeding, partially discussed in the literature, such as the use of a larger-diameter handle in addition to the rotating movement of cutlery (McDonald et al., 2016; Van Room & Steenbergen, 2006; Ma et al., 2008). Also the use of additional weight for food in the case of tremors is mentioned in Hewer et al. (1972), Hömberg et al. (1987), Meshack and Norman (2002), and Ma et al. (2009).

Despite the preliminary results, it is necessary to observe a set of limitations, such as the great variability of results throughout the process, especially with regard to user satisfaction. In this sense, it is worth emphasizing that both performance and satisfaction results depend on the individual's relationship with their occupations and expectations. The individual's family and social context play an important role in these measures, since, according to Law et al. (2009), the environment is composed of physical, social, cultural and institutional elements that influence performance and self-reported satisfaction. In the case of PD, it is also necessary to consider the possible influence of the medication on the user's motor functions on a typical day. Another limitation of this study was the assessment of the gradual filling of AD developed cavities. The single-user method is dependent on the user's health, due to the time needed to infer the results. The weakening of the user's health interrupted the continuity of this specific work.

10.4.1 Test with Volunteers

Ten people (volunteers) with reduced function due to PD and essential tremors were selected from the initial contacts (about 80 e-mails). The academic partnership with the local start-up Model Works allowed the additive manufacturing of ten adaptive eating devices as well as sending them by mail, packaged with a user manual. The inclusion of potential end users for evaluation of the adapted eating device was made possible by an addendum to the previously approved public ethics committee in Brazil (CAAE 35464114.4.0000.5154). The ten selected volunteers received a term of clarification, explaining the conditions and characteristics of the tests. Each of them registered by the design team signed a free consent form. The study with volunteers included sending a kit containing assembled AD for self-feeding, a manual for using the functional prototype and a set of follow-up tables for evaluation during use.

The evaluation process started with the registration in a specific table, for a period of two weeks, of the information on the performance with the conventional cutlery used in the daily routine by the user. After this stage, the use of AD was started, under the following conditions: without water and with one, two, three and four filled cavities. Each configuration was used for two weeks, sequentially, with the results of each one recorded in specific tables. Users' family members monitored the evaluation in some cases and occupational therapists did so in others. The results obtained with this evaluation were not satisfactory. None of the volunteers completed the test cycle recommended by the OTs in the kit, and few even returned the completed tables. Volunteers who responded mentioned interruption of AD use due to difficulties in handling it and lack of face-to-face follow-up by a health care professional.

This preliminary study, without a scientific rigor, brought some important lessons to the design team, especially for engineering members: (1) customized products, mainly in the health area, require, in addition to the product, manual and evaluation means, some type of service guidance and monitoring during use of AD; (2) remote monitoring makes communication even more difficult; (3) there is an exaggerated expectation of the selected potential user, who would not necessarily have a prescription for this type of AD, which ends up discouraging him or her during use, which requires training and persistence of use. In the single-subject tests, it was possible to observe personal involvement of all the people involved in the research project, which helped in the continuity of the study of satisfaction and performance assessments and mainly reinforced the initial adaptation of the user and family to the adapted AD. This should be another aspect that has not been established in testing with volunteers.

This research project has two more partnership developments for the years 2021 and 2022, having paused in 2020 and the beginning of 2021 due to the COVID-19 pandemic: production of a new batch of ten functional prototypes to be evaluated with tests of electromyography with researchers from the *Universidade Federal do Triângulo Mineiro* physiotherapy area and a comparative study of adapted food utensils for people with PD with a focus on ergonomics in partnership with researchers from the Industrial Design area of the Sao Paulo State University, Brazil.

10.5 FINAL CONSIDERATIONS

The main considerations about this research project with the experience acquired in the design development of an assistive device for ADL focused on mass customization were:

- The use of integrated design methods, (HoQ) QFD and TRIZ, especially in the interface of the DCD and conceptualization spiral phases (equivalent to the traditional engineering design phases: technical and economic feasibility and conceptual design), led to more creative thinking by the technical team and allowed the participation of researchers from the health team in collecting information, integrated learning and conceptual solutions. This involvement can be defined as participatory design and user-centered design, with the end user participating in the testing and refinement phase.
- The conceptual phase generated innovative solutions for the AD, and then a utility model patent application (BR2020160243144) was submitted to the Brazilian Institute of Industrial Property (INPI) (Santos et al., 2016).
- Additive manufacturing, through an additive technique and low-cost equipment based on filament extrusion, made it possible to materialize the conceptual solution for discussion among members of the health and technical area until the fabrication and assembly of the technical prototype for tests with end users. Additive manufacturing planning required studies on the combination of process parameters in relation to the improvement of clearances and surface finishing so that, for example, there was confined watertightness in the cavities to carry out tests with end users. The production cost per additive technique of the fourth version of the AD functional prototype for clinical testing was estimated at US$6.00/unit and met the initial proposal for a suitable low-cost AD, including availability for open source design communities.
- Finally, the proposed methodology (AT-8esign methodology) allowed the systematization of interaction and technical iteration processes that are particularly critical in a design with the nature of a customized product, with the appropriate inclusion of stakeholders throughout the project.

ACKNOWLEDGMENTS

The National Council funded this research project for Scientific and Technological Development (CNPq) under process numbers 2016–4/442109. The author thanks all the researchers involved in the project, and particularly the researchers: PhD. Sc.

Santos, A.V. and MSc. Sc. Justino Netto, J.M for supporting the development and improvement of figures and photos presented in this chapter.

REFERENCES

Cavalcanti, A., Amaral, M. F., Silva e Dutra, F. C. M., Santos, A. V. F., Licursi, L. A., & Silveira, Z. C. (2020) Adaptive eating device: Performance and satisfaction of a person with Parkinson's disease. *Canadian Journal of Occupational Therapy,* 87 (3), 211–220.

Chaoqun, D. (2010) Research on application system of integrating QFD and TRIZ. In: *Proceedings of the 7th international conference on innovation & management.* Wuhan University of Technology, China, 499–503.

Chavarriaga, R. et al. (2014) Multidisciplinary design of suitable assistive technologies for motor disabilities in Colombia. In: *IEEE Global humanitarian technology conference (GHTC), 2014. Proceedings . . . IEEE,* 386–391. Infoscience EPFL publishers, Switzerland.

Cheng, L. C., & Melo Filho, L. R. (2007) *QFD: Desdobramento da função qualidade na gestão de desenvolvimento de produtos.* 2a. ed. São Paulo: Edgard Blucher, 539 p.

Clausing, D. (1994) *Total quality development—A step-by-step guide to world-class concurrent engineering.* New York: ASME Press, 506 p.

Coton, J. et al. (2014). Design for disability: Integration of human factor for the design of an electro-mechanical drum stick system. *Procedia CIRP 2014,* 21, 111–116.

De Couvreur, L., & Goosens, R. (2011) Design for (every) one: Co-creation as a bridge between universal design and rehabilitation engineering. *CoDesign,* 7 (2), 107–121.

Deitz, J. C. (2006) Single-subject research. In: G. Kielhofner (Ed.) *Research in occupational therapy: Methods of inquiry for enhancing practice.* Philadelphia: F. A. Davis Company. chap. 11, 140–154.

Forwell, S. J., Copperman, L. F., & Hugos, L. (2008). Doenças neurodegenerativas. In M. V. Radomski & C. A. Trombly-Latham (Eds.) *Terapia ocupacional para disfunções físicas.* São Paulo: Grupo Gen, 1079–1105.

Gadd, K. (2011) *TRIZ for engineers.* West Sussex: John Wiley & Sons.

Gherardini, F. et al. (2018) A co-design method for the additive manufacturing of customised assistive devices for hand pathologies. *Journal of Integrated Design and Process Science,* 22 (1), 21–37.

Harris, D., & Smith, F. J. (2017). What can be done versus what should be done: A critical evaluation of the transfer of human engineering solutions between application domains. In *Engineering psychology and cognitive ergonomics.* Routledge, 339–346.

Hewer, R. L., Cooper, R., & Morgan, M. H. (1972) An investigation into the value of treating intention tremor by weighting the affected limb. *Brain,* 95 (3), 579–590. https://doi.org/10.1093/brain/95.3.579.

Homberg, V., Hefter, H., Reiners, K., & Freund, H. J. (1987) Differential effects of changes in mechanical limb properties on physiological and pathological tremor. *Journal of Neurology, Neurosurgery, and Psychiatry,* 50 (5), 568–579.

Law, M., Cardoso, A. A., Magalhães, S. L. V., & Magalhães, L. C. (2009) *Medida canadense de desempenho ocupacional (COPM).* Belo Horizonte: Editora UFMG.

Liftware. [s. d.] (2015) Available at: <http://store.liftware.com/collections/all/products/liftware-starter-kit>. Accessed on: 28 out. 2015.

Ma, H. I., Hwang, W. J., Chen-Sea, M. J., & Sheu, C. F. (2008) Handle size as a task constraint in spoon-use movement in patients with Parkinson's disease. *Clinical Rehabilitation,* 22 (6), 520–528. http://dx.doi.org/0.1177/0269215507086181

Ma, H. I., Hwang, W. J., Tsai, P. L., & Hsu, Y. W. (2009) The effect of eating utensil weight on functional arm movement in people with Parkinson's disease: A controlled clinical trial. *Clinical Rehabilitation,* 23 (12), 1086–1092. http://dx.doi.org/10.1177/0269215509342334.

Maia, F. N., & Freitas, S. (2014) Proposta de um fluxograma para o processo de desenvolvimento de produtos de Tecnologia Assistiva. *Cadernos de Terapia Ocupacional da Universidade Federal de São Carlos (UFSCar)*, 22 (3).

Majeed, A. et al. (2021) A big data-driven framework for sustainable and smart additive manufacturing. *Robotics and Computer-Integrated Manufacturing*, 67, 102026. ISSN 0736-5845.

Mattazio, R. R. (2019) *Relatório de conclusão de convênio acadêmico*. São Carlos: Modelworks LTDA. 4 p. Relatório técnico apresentado após a conclusão do convênio acadêmico celebrado entre a Universidade de São Paulo e a Modelworks LTDA.

Mayda, M., & Borklu, H. R.(2014) Development of an innovative conceptual design process by using Pahl and Beitz's systematic design, TRIZ and QFD. *Journal of Advanced Mechanical Design, Systems, and Manufacturing*, 8 (3), 1–12.

McDonald, S. S. et al. (2016) Effectiveness of adaptive silverware on range of motion of the hand. *PeerJ*, 4, e1667.

Meshack, R. P., & Norman, K. E. (2002) A randomized controlled trial of the effects of weights on amplitude and frequency of postural hand tremor in people with Parkinson's disease. *Clinical Rehabilitation*, 16 (5), 481–492.

Mihailidis, A., & Polgar, J. M. (2016) Occupational therapy and engineering: Being better together. *Canadian Journal of Occupational Therapy*, 83 (2), 68–69.

Ostuzzi, F. et al. (2015) + TUO project: Low cost 3D printers as helpful tool for small communities with rheumatic diseases. *Rapid Prototyping Journal*, 21 (5), 491–505.

Pachelli, B. L. (2014) *Analysis of the occupational performance of person with Parkinson's disease in the eating task: Conventional spoon x adapted spoon*. Graduation Final Design. Triângulo Mineiro Federal University, Department of Occupational Therapy. 21 p.

Pathak, A., Redmond, J. A., Aallne, M., & Chou, K. L. (2014) A non-invasive handheld assistive device to accommodate essential tremor: A pilot study. *Movement Disorders*, 29 (6), 838–842.

Phillips, B., & Zhao, H. (1993) Predictors of assistive technology abandonment. *Assistive Technology*, 5 (1), 36–45.

Phillips Mobility. *Making life easier*. [s. d.] Disponível em: https://www.tandfonline.com/doi/abs/10.1080/10400435.1993.10132205

Plos, O., Buisine, S., Aoussat, A., Mantelet, F., & Dumas, C. (2012) A universalist strategy for the design of assistive technology. *International Journal of Industrial Ergonomics*, 42 (6), n.6, 533–541.

Portney, L. G., & Watkins, M. P. (2015) Single-subject designs. In: *Foundations of clinical research: Applications to practice*. 3rd ed. Philadelphia: F. A. Davis Company, chap.12, 235–275.

Portnova, A. A. et al. (2018) Design of a 3D-printed, open-source wrist-driven orthosis for individuals with spinal cord injury. *PLoS ONE*, 13 (2), e0193106.

Rossi, A., Berger, K., Chen, H., Leslie, D., Mailaman, R. B., & Huang, X. (2018) Projection of the prevalence of Parkinson's disease in the coming decades: Revisited. *Movement Disorders*, 33 (1), 56–159.

Sabari, J., Stefanov, D. G., Chan, J., Goed, L., & Starr, J. (2019) Adapted eating utensils for people with Parkinson's-related or essential tremor. *The American Journal of Occupational Therapy*, 73, 1–9.

Santos, A. V. F. (2015) *Estudo da melhoria de um utensílio de auxílio à alimentação para portadores da Doença de Parkinson com base na integração das metodologias QFD e TRIZ*. Dissertação de mestrado. Escola de Engenharia de São Carlos, Universidade de São Paulo.

Santos, A. V. F., Cavalcanti, A., & Silveira, Z. C. (2016) *Dispositivo de auxílio para tarefas manuais. BR 20 2016 024314–4 U2*. Depósito junto ao INPI: 18/04/2016. Publicado na Revista de Propriedade Intelectual (RPI) número 2469,

Santos, A. V. F., Licursi, L. A., Amaral, M. F., Cavalcanti, A., & Silveira, Z. C. (2019) User-centered design of a customized assistive device to support eating. *29th CIRP Design 2019—Procedia CIRP*, 84, 743–748. ISSN: 2212-8271.

Santos, A. V. F., Pachelli, B., Cavalcanti, A., & Silveira, Z. (2017) Adaptive design of an eating utensil based on QFD and TRIZ integration. In: *23rd Congress of the European Society of Biomechanics*. Seville, Spain.

Santos, A. V. F., & Silveira, Z. C. (2020) T-design: Methodology to support development of assistive devices focused on user-centered design and 3D technologies. *Journal of the Brazilian Society of Mechanical Sciences and Engineering,* 42, 050202. https://doi.org//10.1007/s40430-020-02347-w.

Schultz-Krohn, W., Foti, D., & Glogoski, C. (2005) Doenças degenerativas do Sistema Nervoso Central. In: L. W. Pefdretti & M. B. Early (Eds.) *Terapia ocupacional: Capacidades praticas para as disfunções físicas*. São Paulo: Roca, 739–764.

Squires, L. A., Williams, N., & Morrison, V. L. (2019) Matching and accepting assistive technology in multiple sclerosis: A focus group study with people with multiple sclerosis, carers and occupational therapists. *Journal of Health Psychology,* 24 (4), 480–494.

Stappers, P. J., Visser, F. S., & Kistemaker, S. (2011) Creation & co: User participation in design. In: B. Van Abel et al. (Eds.) *Open design now*. BIS Publishers, Amsterdam, 140–148.

Stumbo, N. J., Martin, J. K., & Hedrich, B. N. (2009) Assistive technology: Impact on education, employment, and independence of individuals with physical disabilities. *Journal of Vocational Rehabilitation,* 30 (2), 99–110.

Townsend, E., & Polatajko, H. J. (2007) *Enabling occupation II: Advancing an occupation therapy vision for health, well-being & justice through occupation*. Ottawa: Canadian Association of Occupational Therapists.

Tysened, O. B., & Storstein, A. (2017) Epidemiology of Parkinson's disease. *Journal of Neural Transmission*, 124 (8), 901–905. https://doi.org//10.1007/s00702-017-1686-y.

Ullman, D. G. (2010) Designers and design teams. In: *The mechanical design process*, 4th ed. New York: McGraw-Hill, 47–80.

Ulrich, K. T., & Eppinger, S. H. (2008) *Product design and development*. 4th ed. New York: McGraw-Hill.

Van Roon, D., & Steenbergen, B. (2006) The use of ergonomic spoons by people with cerebral palsy: Effects on food spilling and movement kinematics. *Developmental Medicine and Child Neurology*, 48 (11), 888–891.

Vieregge, P., Stolze, H., Klein, C., & Heberlein, I. (1997) Gait quantitation in Parkinson's disease—locomotor disability and correlation to clinical rating scales. *Journal of Neural Transmission*, 104 (2), 237–248.

Zlotin, B., & Zusman, A. (2006) Patterns of evolution: Recent findings on structure and origin. In *Altshuller's TRIZ Institute Conference TRIZCON*.

11 Open Design Concepts (Complementary Text for Chapter 1)

Zilda de Castro Silveira

CONTENTS

11.1 AN OVERVIEW OF THE ENGINEERING PRODUCT

From the handmade artifacts produced by primitive men to the immense variety and complexity of current products, a great framework of technical and scientific knowledge has been and continues to be built. Technology and the intersection of knowledge areas for the improvement and development of new products increasingly demand individual and shared interdisciplinary knowledge. In this context, design techniques can generate the systematization of design activities, and engineering design can be seen in the domain of design science. The engineering product, among its various current applications, with the form tangible of a or non-tangible artifact, is designed and converted to meet society's demands. Boisseau et al. (2018) and Ullman (2010) indicate three fundamental elements to describe product design: "first, the input of the process (the gap); then, the process itself (described through the phases and activities it consists of the boundary objects used, and the stakeholders involved); and, lastly, the output of this process (the plan)". The design of the latter has become a major growing topic (Meroni & Sangiorgi, 2011; Otto & Wood, 2001), notably with the rise of product–service systems. Boisseau et al. (2018) and Ulrich (2011) consider product design a sub-process of product development, where this process is deployed in two sub-processes.

Product design: represents potential conceptual solutions (or a plan) that can meet user needs or social demand (the representation of a gap between expectations and actual need) within a set of multidimensional constraints. This process is oriented toward and performed by design science.

Product manufacturing: represents the structuring of the idea under through physical objects, digital structure or service according to the design plan. It usually involves iteration with the previous step, because its manufacture has a direct impact on the definition of the design product (Boothroyd, 1994; Boisseau et al., 2018). In this way, to obtain a certain artifact, the design process occurs only once, while manufacturing, due to its iterative nature, to reproduce the concept in an assertive way, which can generate several objects. The importance of design activity in the product development process, with its dynamic nature within a multifactorial domain, leads to the structuring of design science or design theory and methodology (DTM). The designation "design science" and product design began to be structured in the second half of the 20th century (Mathias, 2005) in a way that complemented other disciplines, such as management science, business economics and the history of techniques (Boisseau et al., 2018). Yoshikawa (1989) presents a systematic review of design philosophy describing a structure for design science, including historical aspects, which led to a set of practices and methods, as well as the maturity of product development over the years.

The objective was to systematize the experimental technical knowledge acquired over time for product design in order to establish a scientific approach. Therefore, design science is a structuring of engineering design through a combination of theoretical foundations in the areas of science and humanities and recently the incorporation of biomimicry, from natural sciences knowledge, with the set of technical knowledge acquired through empirical applications that generated viable results. Tomiyama et al. (2009) reinforces the use of design theory and methodology in the design process and activities as an iterative approach where the collection of findings supports understanding the different problems levels. In this way, design science is not defined as an exact science because there is a set of uncertainties related to contradictory needs, time and resources, which are subject to the observer's perception but must result in a unique commitment. In this context, design science can be described as an applied science, in which the effective value of a theory or method corresponds to its success in its implementation, which results in a systematic prescription of knowledge (Boisseau et al., 2018).

According to Cross (2007), the engineering design process has been widely studied since the 1960s, mainly focusing on creativity techniques (beginning in the 1950s) and operational research/optimization with management decision-making techniques. The first formal references to design methods, techniques and methodology come with the publication of Assimow in 1962, followed by Alexander (1971), Archer (1965) and the first creativity books with Gordon (1961) and Osborn (1963). Cross (2007) cites definition of "design science" as "an intellectually difficult, analytical, partially formalizable, empirical, and teachable doctrine about the design process". During the 1960s, there was a desire to "scientize" product design.

Cross (2007) described that: "the line of research on engineering design methodology developed strongly in the 1980s, through conferences such as the International Conferences on Engineering Design—ICED". The first structured works started in Germany in the form of technical standards and books on mechanical design guidelines and in Japan (even with limited evidence of practical applications and results). A set of books on design methods and methodology started to be written and published, for example, Hubka (1982), French (1985), Cross (1989) and Pugh (1991). The American Society of Mechanical Engineers (ASME REPORT 1985) launched its series of conferences on "Design Theory and Methodology". Application of DTM during design development, mainly in the early design phase, can add value to the final technical solution, such as minimizing design failures; reducing the total design time and cost; and promoting user participation in order to obtain a list of the main requirements and attributes of the product class, mainly in the conceptual, experimental engineering (tests using functional and technical prototypes) and design parametrization phases.

Books with a focus on industrial product design frameworks represent the main references for DTM cited by Cross (2007). Pahl et al. (2007) represents the semantic school of design philosophy, where the design process can be structured as a technical system that transforms input quantities into output quantities as material, energy and information with the historical technical knowledge acquired over the years. In this context, the authors proposed the systematization of design engineering according to the stages of planning/task clarification, conceptual design and preliminary and detailed design. Each stage has specific activities and expected results, ending with an evaluation process, which guides the transition to the next phase or a new iteration (Pahl et al., 2007). Expanding this idea, Ullman (2010) proposes a model based on six phases: product discovery, product planning, product definition, conceptual design, product development and product support. According to the author, a variety of satisfactory solutions are possible for each problem situation. In this way, the objective of design is to generate a product of quality considering the limitations, time and resource requirements (Ullman, 2010).

Ulrich and Eppinger (2015) proposed a methodology based on six phases: planning, concept development, system-level design, detailed design, testing and refinement and production scale-up. From this perspective, the development process can be seen as an activity of restricting alternatives and increasing specifications, as an information processing system or as a risk management system. It should be noted that these proposed and used engineering design structures were initially conceived for mass and mass customization manufacturing.

Such traditional design views, although already established, were conceived for contexts in which the dominant culture was one of problem solving with cost minimization, whose innovation is driven predominantly by technology through an orderly and linear logical structure (De Couvrerur & Goosens, 2011), where there is no clear interaction with stakeholders, mainly end users in design activities. The traditional design structure created manufacturing paradigms in the mid-19th century. From handmade production, characterized by a great variety and low volume, manufacturing changed, in the beginning of the 20th century, to mass production, historically represented by Henry Ford in the United States and marked by interchangeability, by assembly lines with task division and by a more scientific approach to process management, with time control, training and separation between production and management. The search for productivity with little or no interaction with the user, however, led to a deterioration in the quality of the products and a loss of competitiveness in the face of other products that emerged, especially compared to those from Japan (HU, 2013). In the mid-50s, an industrial philosophy called lean manufacturing emerged based on Toyota's production system, whose main objectives were to maximize consumer value while minimizing waste throughout the manufacturing process.

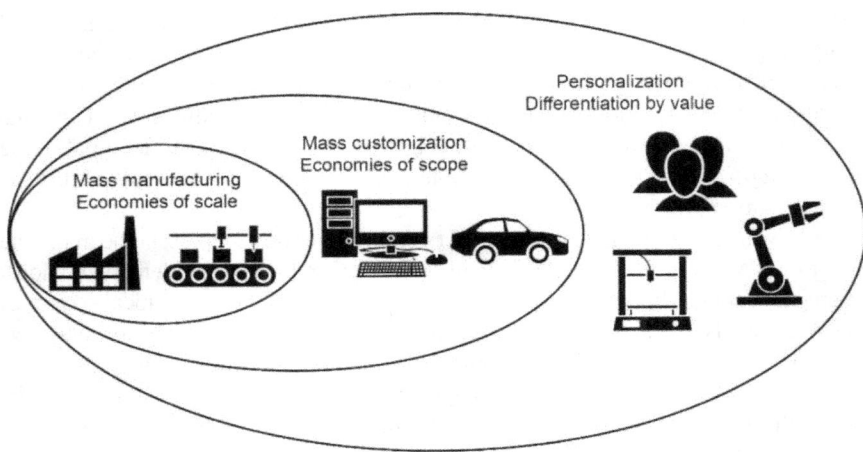

FIGURE 11.1   Manufacturing paradigm goals.

Source: Adapted from Hu, 2013

The paradigm of mass customization emerged in the late 1980s with the increase in demand for the variety of products, having as characteristic features the architecture of product families, reconfigurable manufacturing systems and late product differentiation. Although this paradigm has increased user choice, it has also brought more complexity to assembly systems, influencing their performance. In addition, consumer choice remained restricted to only a limited range of configurations. According to Hu (2013), with the advent of the internet protocol suite or TCP/IP, greater computational capacity for data generation and the maturity of information technology with impact on digital manufacturing provided a new manufacturing paradigm: degrees of customization up to personalization (on-demand manufacturing) as well as design democratization (Hu, 2013; Boisseau et al. (2018). Still, according to Hu (2013), it is possible to differentiate these three paradigms (mass production, mass customization and personalization) into three domains: economies of scale, economies of scope and differentiation of value, respectively.

Rather than being an exclusionary relationship, the relationship between them is one of coexistence (Figure 11.1), in which the newer paradigms encompass the objectives and approaches of the previous ones and establish a demand for more adequate and satisfactory systems. As for the role of consumers/users in each of these paradigms, it can be defined by three distinct verbs, in the following order: "buy", "choose", "design".

11.1.1   Design Democratization and Open Design

Changes in manufacturing paradigms since the First Industrial Revolution until the recent concept of so-called "industrie 4.0" were possible from a set of social and technological changes, mostly driven by information technology and advances in frontier knowledge areas. Regarding the development of the engineering product, there have been several conceptual and technological changes since the incorporation of computer systems in the detailed design and manufacturing phases. Information technology has in a way led to the "democratization" of several other technologies. One of them is related to the emergence of the open-design concept.

Boisseau et al. (2018) points out the impacts of this democratization of design on the descriptive elements of product design: gap/input (co-design in the early stages of design), process (iterative and fractal processes; multidisciplinary stakeholders, collaboration, information reorganization and materialization) and plan/result (set of solutions). These changes brought significant changes to the design process, with the end user's increasingly effective participation in specific phases; digitization of design and manufacturing with the creation of a common language between the two domains of knowledge and, therefore, a need for standardization of product representation and within the context of supervisory systems; and direct machine–machine interfaces. Also according to the authors, the understanding of the concept of open design is recent and still unexplored. Bakirlioglu and Kohatala (2019) report the results of a systematic review of open design in academic areas and conclude that most of the papers reviewed present an empirical nature regardless of the area of knowledge. The authors also highlight that open design emerged as part of the phenomenon of greater participation by citizens, consumers and user institutions and agencies in post-industrial economies. The factors that drive the open design concept are:

- **Digital manufacturing:** in this way, traditional non-traditional manu-facturing processes such as sand-casting, machining and forging, non-traditional changed, and the design process was significantly modified as a function of the virtual environment where a digital design file can move through the entire production chain. The concept of hybrid manufactur-ing with the advances of additive manufacturing in industrial segment has become a reality. On the other hand, additive manufacturing associated with desktop machines is also one of the pillars of the concept of Industry 4.0, driving solutions with low-cost manufacturing by hybrid manufacturing or 3D printing only (Gibson et al., 2015). These changes in manufacturing pro-cesses have led to new forms of production, as mentioned by Boisseau et al. (2018): open production, collective manufacturing and production in pairs, as well as manufacturing as a service.
- **Product digitization:** The second factor is represented by a set of tech-nologies. Computer-aided design (CAD), computer-aided manufacturing (CAM) and computer-aided engineering (CAE) systems were initially stan-dalone with rare computational integrations. Together with management systems such as product life-cycle (PLM), management gradually stand-ardized and started to create a virtual design space. Additive manufacturing technology that can be extended to hybrid manufacturing, including sets of desktop equipment, has only accelerated design democratization.
- **New design structures:** the last accelerator element of open design is related to alternative design structures: fab labs, hackerspaces, makerspaces

and techshops (Cavalcanti, 2013). Fab labs and hackerspaces emerged from the open movement and the creator movement. According to Cavalcanti (2013), these structures are not exactly disruptive. Makerspaces and collaborative development originated from industrial collaborative sites in the 19th century. On the other hand, fab labs (for manufacturing labs), tech shops and hackerspaces are workshops dedicated to personal digital fabrication. According to Boisseau et al. (2018), the main differences are concentrated in relation to volume, degree of user participation and manufacturing technology. The fab lab concept had its start in the university environment (MIT in the early 2000s) at the core of the network, with personal manufacturing machines at an affordable price (Mikahak et al., 2002). Techshops follow the same purposes as fab labs, with the idea of "makerplace franchises" with personal (digital) manufacturing in an open and collective workshop. According to Cavalcanti (2013), "techshop" is a generic term, where personal manufacturing is given as a service and, therefore, has higher rates. Hackerspace refers to a place where a group of programmers share the same physical space (Cavalcanti, 2013). With a focus on computers, they have expanded into electronics and mechatronics projects, which are implemented directly according to the free software movement (open source hardware). Thus, this new design framework has an intrinsic relationship between a digital design approach and manufacturing.

The "open" approach to product design comes from the free software movement. Open source was a movement originally limited to computer engineering, but its pragmatic consequences, the so-called open-x (with "x" as a variable, like DfX, that is, Design for eXcellence), quickly spread to various segments and areas of knowledge, such as open knowledge concepts, open data and open innovation. Bakirlioglu and Kohatala (2019) mention that open design emerged as an intersection of open source software development, DIY maker culture, hacker culture and new understandings of the designer–user relationship. Open design suggests sharing "design knowledge" without limitations and calls for the participation of people with diverse backgrounds to develop iterative design solutions. The design democratization process includes open co-design process, open product development and/or new manufacturing practices and open hardware, and therefore the open design concept promotes social benefits by offering new opportunities for creative processes and open manufacturing, in contrast to passive consumption, where some people influence what is produced.

In the context of the open product, new manufacturing practices arise from the concept of "open design", and the result in the form of a physical object is not yet clear. Therefore, iteration in the construction of conceptual solutions of the idea requires structured participatory methods or co-design, whose idealization can supply the design and manufacturing phases of the product design in design development tasks, such as ideation, decision-making, detailing, prototyping and reflection and review (Bakirlioglu & Kohatala, 2019).

According to Ulrich (2011), product design following two basic elements: the gap, the process itself and the plan, previously mentioned. However, the gap has uncertainties and is independent of the product design: the actors in the design process have no influence on it (that is, on the influence between user expectations and current reality). Thus, the controllable parts of a product design are the plan and the process. Thus, to assess the typology of a product project, two independent axes can be used: the plan and process themselves. Boisseau et al. (2018) present an example of this design space by a mapping, which classifies these relationships in a two-dimensional space: on the abscissa axis, the variable "plan" is identified, ranging from "not open to open", and on the ordinate axis, the variable "process", ranging from not "open to open".

In this product design, space produced different combinations. Considering that the plan is partially open, the process can occur with different natures; starting on the axis of the abscissa with an unopened plan, it is possible to obtain traditional product design, and as it moves significantly along the process axis (ordered), this nature is sequentially transformed into user-centered design, participatory design, open innovation and user innovation. In the same way, crossing from a completely open plan to process results in the concept of open-source innovation to open design. These new product design structures have driven the concept of open manufacturing, bringing the end user/consumer closer to the design, but have also changed the general perception of the industry, which has sought to get closer to its users (Boisseau et al., 2018).

Therefore, Boisseau et al. (2018) define "open-design" as "the state of a design project where both process and the sources of its output are accessible and (re)usable, by anyone and for any purpose".

Open design brought new perspectives on the end user/consumer: stakeholders and end-users appropriating concepts like user-centered design, participatory design, user innovation (Boisseau et al., 2018; Schuler & Namioka, 1993; Norman & Draper, 1986; Stappers et al., 2011 p. 6) and user experience (UX). The user-centered approach considers the user the main actor of the project, establishing ways to look for their real needs. On the other hand, Schuler and Namioka (1993) defined participatory design as "a new approach towards computer systems/software engineering design in which the people destined to use the system play a critical role in designing it".

According to Portnova et al. (2018, p. 10), "users become designers as they offer reflections and suggestions for design improvement based on their experiences and preferences". Stappers et al. (2011) describes the diffuse beginning of co-design, also called pre-design. Indeed, the start of engineering design involves various activities carried out for proper design development, such as the understanding of users and the contexts of use and the exploration and selection of technologies, market opportunities and the niche end-user. After this more critical initial period, the process then moves on to more traditional phases and stages, with the definition of the design criteria, the elaboration of ideas and concepts for the product, the manufacture of prototypes and, finally, obtaining the product.

User innovation is a model proposed by von Hippel and Boisseau et al. (2018), where users are considered the primary source of innovation. In this context, the concepts of co-design or co-creation are included, which refer to "creativity of designers and people not trained in design working together in the design development process" (Stappers et al., 2011, p. 6). The open-design concept offers a set of possibilities, including industrial and local collaborative networks in several product segments: from conception to reuse and replacement. It also drives the concepts of customized and personalized manufacturing. Boisseau et al. (2018) present a scenario of the evolution of product design within the open approach and conclude that the open design typology for tangible artifacts can currently be classified into three categories: do it yourself, meta-design and industrial ecosystems.

- **Do-it-yourself (DIY):** this variation on open design results in evolution and structuring of objects in a very personal and more experimental way. Users share your design, either because they want to share their achievement or because it enables joint works with peers. Richterich (2020) discusses how the DIY communities in the world and in England quickly organized to produce emergencial healthcare equipment, mainly personal protective equipment (PPE) and artificial respiratory valves and systems, due to the COVID-19 pandemic and worldwide collapse in hospital centers. A worldwide open-design network of 3D printing communities (makerspaces and hackerspaces) in different countries employed their skills and tools to produce face shields, surgical masks, respirator valves and mask support. Open design usually refers to products of everyday life that have an analogy to industrial products in order to obtain greater simplicity and lower cost. Desktop equipment, mainly 3D printer–based fused filament fabrication (FFF), were and are thrusters of DIY. Thus, Boisseau et al. (2018) describe that this kind of open design can be defined as a bottom-up initiative, which brings together users' efforts in product development. The motivations are diverse, such as meeting the needs of certain segments, adapting a product based on specific restrictions and reducing product costs. The user's role in product plan and process scales can be to innovate and co-create. This category of open design is quite common in open design communities, such as RepRap (Source: RepRap), Thingiverse (Source: Thingiverse—Digital Projects for Physical Objects) and E-enable (Source: Enabling the Future).
- **Meta-design:** is another variation of open design aligned with the concept of mass customization, where the user can adjust the design according to personal needs (Khalid & Helander, 2003). In this way, the project can be parameterized with objective functions, which include restrictions. According to Boisseau et al. (2018), user access is restricted to a predefined project environment. In this open design approach, it is possible for the user to produce their own design, in modular form, for example. Private companies can use this type of open design through the formation of specialized teams with access to the project restricted to their competences. The products that can benefit from this kind of open design are mass customization and decentralized manufacturing. Some examples are communities: Arduino (Source: Arduino—Home), GitHub (Source: GitHub).
- **Industrial ecosystem:** the last category of open design. In this framework, various stakeholders along the product value chain agree to open up their products and processes. The motivations for adhering to this type of open design are the reduction of development costs and risks, increased process

speed and standardization of the solution to reduce dependence on a monopolistic supplier. This open design category is embedded in the concept of open innovation. An example of this type of open design is the Thin Film Partnership Program, funded by the US National Renewable Energy Laboratory (Source: National Renewable Energy Laboratory [NREL] Home Page | NREL; Buithenhus & Pearce, 2012).

11.1.2 CONSIDERATIONS OF ADDITIVE MANUFACTURING AND ON-DEMAND PRODUCTS

Manufacturing models have adapted to historical demands, often for technology-driven products: from more artisanal manufacturing characterized by high "production" cost, followed by mass production based on interchangeability and custom manufacturing in the 1980s with a focus on scope, giving flexibility and reconfiguration to products (Hu, 2013). However, the most recent concept of design with mass customization has a scope focused on adding value to the product and benefiting from the technology but also including the user as part of the decision processes in the fundamental stages of conceptualization and evaluation of the product.

In the last decade, a new model of design and on-demand manufacturing appeared, driven by the maturation of a set of technologies, including information technology. On-demand manufacturing is based on the individual needs and desires of the end user (by participatory design and user innovation/co-creation), who can participate in the creation design process, interacting effectively with the design team (Hu, 2013). In this process, the product obtained shows high benefit and, sometimes, innovation regardless of the design class. Additive manufacturing (AM) technology contributes to this new design-manufacturing approach and represents one of the pillars of Industry 4.0 (Wohlers et al., 2017).

Design for additive manufacturing (DfAM) presents accelerated growth due to new techniques, often allowing only additive manufacturing to generate flexibility in the product cycle; being able to contribute to the reduction of the lead team; improving the competence of the supply chain with direct manufacturing of sub-assemblies and assemblies; and contributing to reducing material waste, whether due to the nature of construction by layers or by recycling raw material, with less energy consumption, scalable work flow and on-demand manufacturing parts, in addition to eliminating and reducing tooling costs (Leary, 2020). From the point of view of product design, there is no limitation to the geometric configuration combined with the technology of engineering materials that allow a set of structural possibilities and, therefore, applications (Wohlers et al., 2017; Mohammed et al., 2020)

Despite the known advantages, AM has challenges and disadvantages to be overcome related to the adaptation of equipment for mass production while maintaining affordable cost, the difficulty of improving surface finishing without excessive addition to the 3D printing time and the limitation of the available material options, as well as the applicability of AM techniques still showing inherent deficiencies in the process and technical standards still being incipient in order to ensure repeatability and quality for the processes involved (Gao et al., 2015; Galantucci et al., 2015). The concept of construction by two-dimensional (2D) layers to generate a physical object dates back almost 150 years, used at the time to generate free-form topographical maps and photographic sculptures (Bourell et al., 2009). Additive manufacturing as a technology and potential alternative form of manufacturing emerged in the 1980s with the stereo lithography process as documented in the US patent "Apparatus for production of three-dimensional objects by stereo lithography" (Hull, 1986). The filament extrusion process known as FDM (Fused Deposition Modeling, as registered by the company Stratasys) would only appear a few years later with the invention of S. Scott Crump in 1989. The technology has become popular in the last ten years due to the expiration of patents and open-source design communities making replicable and low-cost extrusion 3D printer design available, such as REPRAP and FAB@ HOME (Jones et al., 2011; Malone & Lipson, 2007).

AM is based on the digital environment, with the configuration of a virtual geometric model that can be processed and transmitted on coordinate descriptions. This fact strengthens the process as a pillar of the Fourth Industrial Revolution due to the disruptive characteristic, with great effects on the change in product generation, and contributes to digital fabrication and open design. However, the current scenario is one of research in terms of improving efficiency, reliability and economic viability for the user (Chen et al., 2016). Many geometries need to be adapted to the additive manufacturing model, with prior planning of the parameters that influence the physical properties of the part (i.e. in its mesostructure). The use of support structures and percentage of completion are commonplace examples that show new ways of thinking of the additive system when compared to the subtractive.

The AM techniques are categorized according to the method of structuring and energy applied on the raw material as well as single or multiple steps (ISO/ASTM 52900). In this way, AM extrapolates the concept of being a conventional manufacturing technology (Kumke et al., 2016), as it generates prototypes, components and products from the addition of layers/voxels, with minimal waste of material and reduction of tooling for post-processing. Kumke et al. (2016) also mentions that additive manufacturing allows innovative designs, free of predefined shapes and parts with estimated performance. The possibility of manufacturing products with a high degree of personalization (on-demand product or mass customization) is widely referred to as the concept of Industry 4.0 (Wahlster et al., 2014).

In the context of maturation of AM, adaptable solutions are desired to facilitate the iteration with the user. The term "smart additive manufacturing" is used for the integration of MA systems in order to respond in real time to end-user needs (Verboeket & Krikke, 2019; Daduna, 2019; Wang et al., 2020; Majeed et al., 2021). In addition to the availability of networked files, which is already a current reality, a scenario is envisaged in which artificial intelligence can alter print files in real time (Yang et al., 2017). Additive manufacturing goes beyond the unique concept of being an unconventional manufacturing process, since it brings together knowledge and technologies from different engineering areas, with an interface in health areas. In the health area, in addition to applications in the area of tissue engineering, orthotics, prostheses, bone engineering and surgical tools, the manufacture of delivery drugs has become a fruitful field of application, where it has become possible to customize drug fractions by manipulating the gradual release and drugs in order to provide a specific therapeutic effect (Mohammed et al., 2020). However, online and online quality control and time are still conditions that limit the industrial-scale use of drug manufacturing with additive processes.

Ghilan et al. (2020) present variations related to the term "3D printing" established in mid-1984 from the interactions between areas of knowledge: As of 2010, the technological improvement of additive processes started to include engineering materials and mechatronics items, as well as the increased demand in multidisciplinary areas giving rise to variations in 3D printing, such as 3D bioprinting (2010), 4D printing (2013) and 4D bioprinting (2014), a specialized extension of 3D bioprinting that aims to reconstruct biochemical and biophysical composition, as well as the hierarchical morphology of various tissues using stimuli-responsive biomaterials and cells, fundamental characteristics in pharmaceutical and medical research cited in personalized therapy and five-dimensional (5D) printing (2016). Willian Yerazunis's Mitsubishi Electric Research Laboratories (MERL) introduced five-dimensional printing in 2016. Five-axis 3D printing is an extension of 3D printing in which the print head has the ability to move in five degrees of freedom due to a mobile construction area (Deitz et al., 2006).

The concept of design for additive manufacturing derives from Design for eXcellence and can be briefly defined as an integrated approach based on additive manufacturing, interdisciplinary and multidisciplinary procedures and methods still under development (Thompson et al., 2016). This new technological paradigm, provided by DfAM, one of the pillars of the Industry 4.0 concept, requires a change in industrial culture, which has already been noticed in the open design and open manufacturing communities. However, there is a difficult path to overcome the "cognitive barriers" imposed by previous experiences and conventional manufacturing techniques imposed by industry and academia. Readjustments in engineering education, as well as changes in engineering, are becoming more and more necessary. Thus, design science, with its structured design methods, represents an important element in the context of a virtually digital environment.

Finally, De Couvreur and Goosens (2011, p. 113) define co-design as "a set of iterative techniques and approaches that place users in a prominent role in tasks and phases of product design". From industrial products, until healthcare products are designed on demand, there is a need for design customization. The union of co-design with DfAM presents the potential to promote a participatory design environment, counting on the evolution from conceptual prototype to functional/technical prototype materializing.

REFERENCES

Alexander, C. (1971) The state of the art in design methods. *DMG Newsletter,* 5(3), 3–7. Cliffs, NJ.

Archer, L. B. (1965) *Systematic Method for Designers.* The Design Council, London.

ASME REPORT. (1985) *Goals and Priorities for Research on Design Theory and Methodology.* National Science Foundation.

Bakirlioglu, Y., & Kohatala, C. (2019) Framing open design threough theoretical concepts and practical applications: A systematic literature review. *Human-Computer Interactions,* 0, 1–44. Taylor & Francis. ISSN 0737-0024.

Boisseau, E., Omhover, J-F., & Bouchard, C. (2018) Open-design: A state of the art review. *Design Science Journal,* v(4), 3.

Bourell, D. L., Beaman, J. J., Leu, M. C., & Rosen, D. W. (2009) A brief history of additive manufacturing and the 2009 roadmap for additive manufacturing: Looking back and looking ahead. *Proceedings of RapidTech,* 24–25.

Buithenhus, A. J., & Pearce, J. M. (2012) Open-source development of solar photovoltaic technology. *Energy for Sustainable Development,* 16, 379–388.

Cavalcanti, G. (2013) Is it hasckerspace, makerspace, techshop or fablab? *Make,* 22. *Is it a Hackerspace, Makerspace, TechShop, or FabLab? | Make: (makezine.com).* Access: 05/05/2021.

Chen, A. et al. (2016) Reprise: A design tool for specifying, generating, and customizing 3D printable adaptations on everyday objects. *UIST,* 29–39, October.

Cross, N. (1989) *Engineering Design Methods.* John Wiley & Sons Ltd, Chichester.

Cross, N. (2007) Forty years of design research. *Design Studies,* 28(1), 1–4.

Daduna, J. R. (2019) Disruptive effects on logistics processes by additive manufacturing. *IFAC-PapersOnLine,* 52(13), 2770–2775. ISSN 2405-8963. *9th IFAC Conference on Manufacturing Modelling,* Management and Control MIM 2019.

De Couvreur, L., & Goosens, R. (2011) Design for (every) one: Co-creation as a bridge between universal design and rehabilitation engineering. *CoDesign,* 7(2), 107–121.

Deitz, J. C. (2006) Single-subject research. In: Kielhofner, G. *Research in Occupational Therapy: Methods of Inquiry for Enhancing Practice.* F. A. Davis Company, Philadelphia, chap. 11, 140–154.

French, M. J. (1985) *Conceptual Design for Engineers.* The Design Council, London

Galantucci, L. M., Guerra, M. G., Dassisti, M., & Lavecchia, F. (2019, June). Additive Manufacturing: New Trends in the 4th Industrial Revolution. In *International Conference on the Industry 4.0 model for Advanced Manufacturing* (pp. 153–169). Springer, Cham.

Gao, W. et al. (2015) The status, challenges, and future of additive manufacturing in engineering. *Computer-aided Design,* 69, 65–89.

Ghilan, A., Chiriac, A., Nita, L., Rusu, A., & Neamtu, I. (2020) Trends in 3D printing processes for biomedical fields: Opportunities and challenges. *Journal of Polymers and the Environment,* 28, 1345–1367.

Gibson, I., Rosen, D., & Stucker, B. (2015) *Additive Manufacturing Technologies: 3D Printing, Rapid Prototyping and Direct Digital Manufacturing.* 2nd ed. Springer, New York, 498 p.

Gordon, W. J. J. (1961) *Synectics.* Harper & Row, New York.

Hu, S. J. (2013) Evolving paradigms of manufacturing: From mass production to mass customization and personalization. *Procedia Cirp,* [s.l.], 7, 3–8.

Hubka, V. (1982) *Principles of Engineering Design.* Butterworth Scientific Press, Guildford.

Hull, C. (1986) *Apparatus for Production of Three-Dimensional Objects by Stereolithography.* US4575330.

Jones, R. et al. (2011) REPRAP—the replicating rapid prototype. *Robotica,* 29, 177–191, 01 201.

Khalid, H. M., & Helander, M. G. (2003) Web-based do-it-yourself product design. In: M. M. Tseng & F. T. Piller (Eds.) *The Customer Centric Enterprise: Advances in Mass Customization and Personalization.* Springer, New York, 247–266.

Kumke, M., Watschke, H., & Vietor, T. (2016) A new methodological framework for design for additive manufacturing. *Virtual and Physical Prototyping.* 11(1), 3–19.

Leary, M. (2020) *Design for Additive Manufacturing.* Elsevier. [s.n.]. ISBN 978-0-12-816721-2.

Majeed, A. et al. (2021) A big data-driven framework for sustainable and smart additive manufacturing. *Robotics and Computer-Integrated Manufacturing,* 67, 102026. ISSN 0736-5845.

Malone, E., & Lipson, H. (2007) Fabhome: The personal desktop fabricator kit. *Rapid Prototyping Journal,* 13, 245–255, 08.

Mathias, H. (2005) *Kunst und Wissenschaft in der Technik des 20.* Zur Geschichte der Konstruktionswissenschaft. Chornos, Jahrhunderts.

Meroni, A., & Sangiorgi, D. (2011) *Design for Services. Design for Social Responsabilities.* Gower Publishing, Ltd., Burlington, VT.

Mikahak, B., & Lyon, C. (2002) Fab lab: An alternate model of ICT for development. In: *Second International Conference an Open Collaborative Design for Sustainable Innovation*. Thinkcycle, Bangalore.

Mohammed, A., Elshaer, A., Sareh, P., Elsayed, M., & Hassanin, H. (2020) Additive manufacturing technologies for drug delivery applications. *International Journal of Pharmaceutics*, 580, 119245. https://doi.org/10.1016/j.ijpharm.2020.119245

Norman, D. A., & Draper, S. W. (1986) *User Centered System Design. New Perspectives on Human-Computer Interaction*. CRC Press, Boca Raton, FL.

Osborn, A. F. (1963) *Applied Imagination pp Principles and Procedures of Creative Thinking*. Scribner's Sons, New York.

Otto, K., & Wood, K. (2001) *Product Design: Techniques in Reverse Engineering and New Product Development*. Prentice Hall, New York.

Pahl, G., Beitz, W., Feldhusen, J., & Grote, K. H. (2007) *Engineering Design: A Systematic Approach*. 3rd ed. Springer-Verlag, Berlin.

Portnova, A. A., Mukherjee, G., Peters, K. M., Yamane, A., & Steele, K. M. (2018) Design of a 3D-printed, open-source wrist-driven orthosis for individuals with spinal cord injury. *PloS one*, 13(2), e0193106.

Pugh, S. (1991) *Total Design: Integrated Methods for Successful Product Engineering*. Addison-Wesley, Wokingham.

Richterich, A. (2020) When open source design is vital: Critical making of DIY healthcare equipment duirng the COVID-19 pandemic. *Health Sociology Review,* 29(2), 158–167. Taylor & Francis.

Schuler, D., & Namioka, A. (1993) *Participatory Design: Principles and Practices*. Lawrence Erlbaum Associates, Hillsdale, NJ.

Stappers, P. J., Visser, F. S., & Kistemaker, S. (2011) Creation & co: User participation in design. In: B. Van Abel et al. (Eds.) *Open Design Now*. BIS Publishers, Amsterdam, 140–148.

Thompson, M. K., Moroni, G., Vaneker, T., Fadel, G., Campbell, R. I., Gibson, I., Bernard, A., Schilz, J., Graf, P., Ahuja, B., & Martina, F. (2016) Design for additive manufacturing: Trends, opportunities, considerations and constraints. *CIRP Annnals—Manufacturing Technology,* 65, 737–760.

Tomiyana, T., Gu, P. et al. (2009) Design methodologies: Industrial and educational applications. *CIRP Annals—Manufacturing Technology*, 58(2), 543–565.

Ullman, D. G. (2010) Designers and design teams. In: *The Mechanical Design Process*. 4th ed. McGraw-Hill, New York, 47–80.

Ulrich, K. T. (2011) *Design: Creation of Artefacts in Society*. 1st ed. University of Pennsylvania, Philadelphia.

Ulrich, K. T., & Eppinger, S. H. (2015) *Product Design and Development*. 4th ed. McGraw-Hill, New York.

Verboeket, V., & Krikke, H. (2019) The disruptive influence of additive manufacturing on supply chains: A literature study, conceptual framework and research agenda. *Computers in Industry*, 111, 91–107. ISSN 0166-3615.

Wahlster, W., Gallert, H.-J., Wess, F. H., & Widenka, T. (2014) *Towards the Internet of Services: The Theseus Research Program*. Springer, New York. ISBN 978-3-319-06755.

Wang, Y. et al. (2020) Smart additive manufacturing: Current artificial intelligence-enabled methods and future perspectives. *Science China Technological Sciences,* 63(9), 1600–1611.

Wohlers, T. et al. (2017) *Wohlers Report 2017: 3D Printing and Additive Manufacturing State of the Industry*. Fort Collins: Wohlers Associates, 344 p.

Yang, J., Chen, Y., Huang, W., & Li, Y. (2017, September). Survey on artificial intelligence for additive manufacturing. In *2017 23rd International Conference on Automation and Computing (ICAC)* (pp. 1–6). IEEE.

Yoshikawa, H. (1989) Design philosophy: The state of the art. *Annals of the CIRP*, 38(2), 579–586.

Index

For Product Safety Concerns and Information please contact our EU
representative GPSR@taylorandfrancis.com
Taylor & Francis Verlag GmbH, Kaufingerstraße 24, 80331 München, Germany

www.ingramcontent.com/pod-product-compliance
Lightning Source LLC
Chambersburg PA
CBHW071414180526
45170CB00001B/99